WHAT PEOPLE ARE SAYING ABOUT

Mental Penguins

I literally could not put this book down. Prof. Sardamov makes a passionate, meticulously-researched and utterly compelling case for reinstating reading (yes: old-fashioned text-based reading) at the heart of formal education. No UK or US academic at the moment would dare to write this book—but, boy, do we need it! Buy it, read it and send a copy to your favourite politician.
Sue Palmer, literacy specialist, author of *Toxic Childhood* and *Upstart*

Prof. Sardamov sounds the alarm about the unrelenting, pervasive stimulation facilitated by our enchantment with information technology. He makes a compelling case for reclaiming the less thrilling yet essential gratifications of reading, one child at a time.
Dr. Philip Zimbardo, Professor Emeritus, Stanford University, and Founder of The Heroic Imagination Project

An impassioned plea for the need to avoid gimmickry in education and to recapture the patient reading and learning that gives depth and breadth to developing minds. Those who have their own reasons not to hear its important message will easily dismiss this very personal book, but they would be unwise to do so.
Dr. Iain McGilchrist, psychiatrist, author of *The Master and His Emissary*

Prof. Sardamov has accomplished a fabulous integration of the personal narrative and academic form, with a readable, understandable call to alarm for anybody willing to listen. His own

experience is compelling, and his review of many aspects of neuroscience, psychology and even philosophy lend tremendous theoretical support to his argument.
Dr. Stephanie Brown, clinical psychologist, author of *SPEED*

Mental Penguins

The Neverending Education Crisis and the False Promise of the Information Age

Mental Penguins

The Neverending Education Crisis and the False Promise of the Information Age

Ivelin Sardamov

BOOKS

Winchester, UK
Washington, USA

First published by iff Books, 2017
iff Books is an imprint of John Hunt Publishing Ltd., Laurel House, Station Approach,
Alresford, Hants, SO24 9JH, UK
office1@jhpbooks.net
www.johnhuntpublishing.com
www.iff-books.com

For distributor details and how to order please visit the 'Ordering' section on our website.

ISBN: 978 1 78535 342 0
978 1 78535 343 7 (ebook)
Library of Congress Control Number: 2016939770

A CIP catalogue record for this book is available from the British Library.

Design: Stuart Davies

Printed and bound by CPI Group (UK) Ltd, Croydon, CR0 4YY, UK

CONTENTS

It would be all too comical
Wasn't it so deeply sad.[1]
—M. Y. Lermontov

1

The End of Education Revisited

This is the story of how I solved a mystery—the mystery of the neverending education crisis. Not that I have found any ultimate solution or set of policy and pedagogical fixes for it—that would be too much to ask. Rather, I believe I have come to understand why so much hope and hype, effort, and resources invested in education and educational reforms have failed to produce generations of better prepared high school and college students—not just in the United States, but also in many other countries. And why, as they have adapted to a harsh and dynamic sociotechnological environment, growing numbers of students seem unimpressed by the ideas and the larger social world they are required to study (despite occasional bouts of enthusiasm like the wave of youthful support for American presidential hopeful Bernie Sanders or other causes).

Alas, I have come to the conclusion that the most "we" — those apprehensive about the socio-technological challenges I will describe—can hardly turn the tide. The most we can do is try to construct small-scale sanctuaries for ourselves, our students, and our children. This must be a depressing conclusion, but it seems unavoidable (to me at least) given the societal and technological pressure cooker in which we live—and in which we need to raise our children.

To reach this conclusion, I have stood on the shoulders of countless neuroscientists and other scholars—and perhaps stepped on the toes of many. I have also benefited from observing my own mental Galapagos (or Antarctica)—the assortment of students I have taught at the American University in Bulgaria over the past 18 years. For the better part of these years, I have been puzzled by a central paradox. As I have worked to become a better college teacher, I have seen growing numbers of my students struggle to grasp the main points in course readings, and to accumulate knowledge and connect the dots as they take various courses. And I have observed growing estrangement from larger social issues and ideas, and confusion even about simple administrative policies and requirements.

Of course, these concerns are not entirely new. "The Crisis in Education" was, in fact, the title famed political philosopher Hannah Arendt chose for an article she published over six decades ago.[1] It offered a somewhat alarmist analysis at a time which some traditionalists still see as a period of solid learning in American schools and colleges. By the 1950s, however, perceptions of a far-reaching crisis in American education had already become conventional wisdom. Seeking an explanation and a way forward, Arendt lamented what she saw as the crass instrumentalization of education—its reduction mostly to a tool for achieving other goals. She thought that approach was counterproductive, even when pursued with an ostensibly noble purpose—the cultivation of engaged and responsible citizens.

Arendt wrote about the area of education which was closest to her heart (and resonates most in mine)—the teaching of political ideas. But the problems she addressed ran much deeper. As a political philosopher, Arendt understood well the need to place these problems in a broader social and intellectual context. She was concerned that the problematic approach to education she was criticizing in fact reflected and was reinforcing a more comprehensive crisis of modern society—a society which needs to find a balance between attachment to healthy traditions and the demands of the future.

Arendt's criticisms were directed not only at the still predominant "factory" model of education. She also believed the "progressive"[2] educational theories and practices (associated with John Dewey and other educational reformers) overturned too radically all pedagogical tradition. The 1960s, however, saw the explosive growth of ever more "progressive" educational models. One of the many to welcome the ostensibly subversive promise of this trend was media and education guru Neil Postman.[3] In the late 1970s, however, he already saw a need for teaching to become a "conserving activity."[4] He hoped such an approach could counteract disturbing cultural and technological trends, and a potentially debilitating unbalancing of the information environment.

Postman was initially most concerned with the ubiquity of television, the predominance of visual imagery, and the danger of "amusing ourselves to death ... in the age of show business."[5] The spread of personal computers later convinced him that the allure of visual images could have fatal cultural and educational consequences even outside their use in conventional entertainment. In the mid-1990s he was already lamenting "the end of education,"[6] marked by the demise of the grand narratives that could endow education with a larger purpose. Both Arendt and Postman hoped that the troubling trends they had identified could be corrected—by returning to a less utilitarian model or the spread of some "noble lies" regarding the larger goals of education.

I am afraid my own vision is not as cautiously optimistic as theirs—or as good humored as Postman's. Worse, my skepticism has been reinforced[7] by a torrent of books which have offered a spirited defense of liberal education and the role of the humanities in it, increasingly seen as facing a grim future.[8] At the same time, many hard-headed reformists have gone in the opposite direction in primary and secondary education.[9] Unburdened by any larger philosophical perspective, they have often focused on classroom techniques, educational technologies, and algorithmic data analysis. They have also designed strategies for taking over the commanding (and conceptual) heights of education from the old guard. Of course, this game plan reflects precisely the instrumentalist fallacy Arendt once sought to expose in the sphere of education and beyond.

These debates have been focused not only on educational practices but also on wars of ideas and political projects. From my perspective, both the philosophical probings of humanistic educators and educational theorists and the no-nonsense recipes of their utilitarian opponents miss a vital dimension. They do not sufficiently take into proper consideration the intricate ways in which the social and mental lives of children and adults are enmeshed with some basic neural and physiological processes. This statement will probably evoke an instant rebuttal pointing to the recent mushrooming of neuroeducation,[10] alongside many other neuro-branded fields of research. From my investigations, however, I have concluded that these undertakings have some serious limitations.

Neuroscience has certainly produced captivating images of the human brain and many curious findings regarding the neural "correlates" of mental processes. These results have been supplemented by clever experiments in "social neuroscience" demonstrating that the configurations of neural networks, in their turn, can be influenced by specific stimuli within our social and technological environment. Most neuroscientists, though, have sought to produce fairly mechanistic explanations identifying causal relationships between isolated neural and behavioral or environmental variables. They therefore open themselves to the accusation that they offer reductionist and deterministic accounts.[11] The larger concern about neuroscientific causal models is that, once disseminated in simplified form by science journalists, they can undermine our indispensable faith in free will and personal responsibility.[12]

Such worries may have a point. But my sense is that notions of free will and individual agency have been undermined not by irresponsible theorizing but rather by much larger social and technological forces—which have also twisted the objectives and outcomes of education at all levels. I suspect that ignoring these forces, and seeking to bracket the potentially troubling implications of some recent findings in neuroscience (the "let's pretend" strategy), can create its own problems. It can inspire misplaced faith in different varieties of intellectual or moral proselytizing, to say nothing of the wishful thinking underlying the growing reliance on dubious abstract models and technical solutions—from the unyielding devotion of economists to the alleged efficiency of financial markets and the commodification of public services and spaces, to the recent avalanche of distance learning start-ups and schemes.

My own approach will, no doubt, provoke accusations both of crass reductionism and of making willful inferences unsupported by reliable experimental or empirical data. I am not sure I can do much to dispel such doubts. My strategy in confronting them will be to offer a broad picture that weaves together the complex patterns of interplay between processes at the level of neural and other cells, the skull- and skin-bound human organism, and the human anthill—including the increasingly pervasive presence of information and other technologies in our lives.

Alas, this is easier said than done. No matter how complex and nuanced the picture that emerges from my musings, it is bound to have many deficiencies and blind spots. The extensive references I provide may also appear less than convincing to critics with a very different mindset and selection bias. I do anticipate such criticisms since I am very much aware that the picture I draw will be influenced by some personal and cultural idiosyncrasies. For starters, I grew up in Bulgaria—a country proclaimed by *The Economist* as the unhappiest place on Earth as proportionate to its GDP per capita[13] (a distinction it has achieved in many other surveys of self-reported happiness or life satisfaction). I later went on to earn an American Ph.D. in political science, but resisted the pursuit of scientific rigor which had come to dominate the discipline. Instead, I have become deeply influenced by some broader Western intellectual traditions. I have also followed selectively the intellectual debates covered by quality publications aimed at the American liberal intelligentsia (like *The New York Times, The New York Review of Books, The Atlantic,* and a few others).

Exposure to this very different existential and intellectual ether has helped me put in perspective my somewhat fatalistic Bulgarian dispositions. But I have been unable, and partly unwilling, to completely overcome these. Speaking at an academic conference on political science education, I once referred (in the conventional self-deprecating way) to the strong pessimistic streak in Bulgarian culture. I did this because I felt like the designated Cassandra on a panel brimming with enthusiasm for innovative approaches involving experiential and distance learning. The chair of the panel commended me half-jokingly for being a great representative of my country—and I almost took it as a compliment.

Trying to perhaps make a virtue out of necessity, I have sometimes compared my own predicament to that of Alexis de Tocqueville. He was well aware of his own aristocratic biases, but could not quite escape their grip. In a letter he wrote as he was traveling around the United States, he confessed as much:

> Bound to the royalists as I am by a few common principles and a thousand family ties, I find myself in a way chained to a party whose conduct often strikes me as dishonorable and almost always extravagant. I can't help taking their faults deeply to heart, even as I condemn them with all my might.[14]

There was maybe one peculiar reason why Tocqueville did not seek to fully transcend the narrow-mindedness of the French aristocracy. He recognized that this point of departure opened his senses to some social, cultural and psychological dimensions of

American life in which the natives were swimming like the proverbial fish in water. He believed there were "certain truths which the Americans can learn only from strangers,"[15] and he gladly took this role of a prophet dispatched to a foreign land.

Tocqueville's ambition, however, was much broader. He sought to identify in American society aspects of what he saw as the future shape and character of the civilized world, once deep class divisions had become obsolete. The "equality of condition" he observed in the United States impressed him as a clear improvement on the staleness and injustices of France's semi-medieval society. Yet he observed some troubling tendencies in American society which would soon have larger implications for the social upheaval sweeping across Europe.

I am well aware that the American reading public must be sick and tired of Tocqueville epigones like French philosopher Jean Baudrillard and British literary theorist Terry Eagleton.[16] So I would not claim that I can in any way approach Tocqueville's intellectual stature and perceptiveness. Rather, I look up to him as a cultural hero and a source of inspiration. Of course, a degree of experiential dissonance is not a straightforward avenue to illumination. I have now very much resigned myself to being stranded between two very different social, cultural, and intellectual worlds I inhabit. I also recognize that many of the conclusions I reach departing from this existential posture may come across as awkward.

This impression will probably be reinforced by my writing style. Part of its peculiarity comes from the fact that I am not a native speaker. Though I have mastered the basics of English grammar and syntax, I can never mimic the narrative flair of a Joseph Conrad or even a Joseph Brodsky. Recognizing these limitations, I have sought to avoid complex syntax[17] and any attempts at literary flourish. If my text still comes across as somewhat dense and unwieldy, this in itself may convey a poetic truth. In fact, I would not mind sounding like a somewhat morose and didactic Eastern European intellectual bent on sweeping generalizations—as long as I can avoid any immediate association with Slavoj Žižek.[18]

In addition to some broader influences, I do bring into my investigation a degree of personal and intellectual eccentricity. I have always had some brooding tendencies which may seem a bit excessive even by Bulgarian standards. My inclination to usually see the glass as half-empty and to hear (and ring) cultural alarm bells should thus be taken with a grain of salt. On the other hand, a dour outlook may have some unexpected upsides. Depressive predilections have recently been associated with flashes of creative insight and an ability to associate issues and ideas which seem unrelated.[19] So I do not display much false modesty when I am occasionally praised for my "lateral thinking." I have in fact come to consider this as my intellectual "comparative advantage." This is, in fact, the self-flattering justification I have embraced for shunning empirical research and drawing instead on the hard work of others[20]—unless some of the associations I draw are, indeed, overly far-fetched.

These detailed, slightly defensive self-revelations may appear as overly self-absorbed in the introduction to a book trying to set heftier issues in a broader context. I did feel a need, however, to include them as a caveat at the outset of what is a deeply personal account of the way in which I have moved from one insight to the next, reaching some unsettling conclusions. I suspect I may have lost a few readers at this point. Nor am I sure how many will be willing to remain plugged into the mental matrix I seek to weave to the end. My overall skepticism has helped dispel any illusions that I can win over throngs of converts and help make the world a better place.

In fact, I seriously doubt that I can convince anyone who, when they place their hand on their heart and look around and into the eyes of children and adolescents, does not share at least part of my anxieties and premonitions. There is much recent research which indicates that the extent to which we embrace ideas, political platforms, and even political candidates depends largely on some unconscious emotional responses, which then trigger trains of rationalizations.[21] If this is the case, I should not harbor the illusion that I can win over those who do not cringe as easily at the cultural and technological developments I will describe.

With this in mind, I have chosen to address my argument to the small subset of potential readers for whom my observations may have some emotional resonance. This means that my book is bound to lack broad commercial appeal.[22] It is also devoid of strict "scientific" pretensions. And though I have strong confidence in my overall conclusions, I am willing to entertain criticisms.

With these caveats regarding my intellectual, cultural, and personal perspective, I now stand ready to present my theory of the perennial crisis in education. It could initially be summarized in three main points: 1) chronic dopamine overload (induced by technological saturation and sensory overstimulation), which has contributed to 2) warped brain plasticity, and thus 3) a crisis of curiosity, motivation, and implicit learning in the majority of students, and growing utilitarian detachment in "the best and the brightest."

As these points may seem a bit crude and schematic, I have also tried to place the trends they capture in a broader social context. If my theory still appears outlandish or overly alarmist, your optimal decision would probably be to put this book aside. If you are not completely sure what I have in mind, or you can vaguely anticipate my argument and you still want to give it the benefit of the doubt, please read on.[23]

As I promised, I will trace back the personal journey which has taken me to some disheartening conclusions—which I would rather have avoided. If at some point you feel my attempts to uncover interlocking patterns in the neural and sociotechnological tendencies I describe have lost relevance, you can easily shrug these off and move on to a more reassuring read. If you do decide to stay with me at least provisionally, but would like to move along faster, you can initially ignore the personal vignettes and diversions I will include in notes and the boxes I have posted throughout the book.[24] You may or may not decide to return to these later.

If I can reach across and help a few kindred souls acquire a richer grasp of their own—and their children's—predicament, the time and energy needed to complete this book will be well spent. I would be particularly gratified if some of my current and former students can relate to my thoughts. If even these humble goals are overly ambitious, I still feel a need to collect my scattered thoughts in a "codex" which is part self-discovery, part self-help, and completely honest—for myself and anyone whose heart I can touch.

The most important member of my reading audience will, of course, be our teenage daughter Gali. She is the one person who must find this book personally significant—just because her dad wrote it.

2

College Teaching and Its Discontents

It should come as little surprise that all my adult life I have been nagged by a fuzzy sensation that the time we live in is once again out of joint. I have consequently been attracted to a wide variety of skeptical and alarmist social theories—from Rousseau's argument that the ostensible progress of the arts and sciences had resulted in moral corruption, through Marcuse's critique of consumer capitalism, to conservative or communitarian laments that the Protestant ethic, respect for authority, and a degree of social cohesion have been displaced by a culture of impulsive self-indulgence. What set me on a clearer intellectual course, though, were the problems I began to notice as a college teacher over a decade ago.

When I started teaching political science at the American University in Bulgaria (AUBG) in the fall of 1998, I was not the greatest teacher. I was fresh out of graduate school at the University of Notre Dame. Its still very traditional Department of Government and International Studies[1] had provided me with an intellectual environment that was at once cozy and stimulating. There, I had been inspired by a few great teachers like Fred Dallmayr, Alan Dowty, and Martha Merritt—but had received zero pedagogical training. I had asked to audit a course on teaching methods, only to find out that the professor offering it taught mostly by negative example. Class materials and sessions were mildly disorganized, and she was often late for class. I had also taught my own course on nationalism, but no one from the department had given me any guidance or shown any interest in it. All this experience was hardly sufficient preparation for facing the college classroom.

The courses I taught at AUBG were initially based on a few

fairly dull textbooks. In upper-level classes, I supplemented those with sometimes duller 30-page journal articles rehearsing the main theoretical debates in different areas. Naturally introverted, I did not exactly feel the classroom as my natural element. To reduce my anxiety, I stuck to carefully laid out lesson plans. I also presented detailed outlines of assigned readings analyzing, for example, World Bank policies or the roots of ethnic conflict. As a nod to the interactive teaching methods AUBG faculty were encouraged to adopt, I did direct occasional questions at students. But against much common advice, I often went on to answer these myself, and the classroom rarely erupted in heated discussion.

My initial approach to teaching thus violated all postulates of pedagogical wisdom which, ever since Dewey, had increasingly encouraged active student engagement. Yet, something almost miraculous happened. Most of the exams I graded showed solid understanding of difficult concepts, reasonable clarity of thought, and a decent command of written English. I also saw quite a few students, most of whom never received an A, sign up to take three of four courses with me. One of them, Denitsa, even came by (after I had given her yet another B+ on her take-home final essay) to tell me how much she and her friends loved my courses. She said they joked they were majoring in "Sardamov," not in political science.

I was slightly puzzled by such professed or implicit student devotion which I could not possibly have inspired. I was even more perplexed by the ability of most of my students to demonstrate some solid learning under such suboptimal conditions and the cognitive torture to which I was subjecting them. Despite those encouraging, if surprising, results, I sensed I needed to liven up my classes a bit. I gradually replaced many unwieldy journal articles with shorter, less "theoretical," and more digestible texts from publications like *Foreign Affairs*. I also introduced a larger number of sometimes longer, but much more evocative pieces like Robert Kaplan's dark prophesies from *The Atlantic*.[2]

I also started to open each class session by asking students about current events covered in the media. I additionally integrated into my classes short news stories or opinion pieces which I handed out for students to read in class. Those would typically contain lively descriptions of momentous events like the Asian financial crisis or NATO's bombing campaign over Serbia. Both our current-event roundups and the handouts I distributed often supplied striking details from the unending real-life drama ostensibly captured by dry theoretical debates. These were meant to provide vivid illustrations of the larger issues and more abstract concepts discussed in course readings.

This purpose was served even better, I thought, by brief segments from recent documentaries (mostly produced by the BBC) which I started to show regularly in class. For example, I would incorporate into a class dealing with nationalism and ethnic conflict interviews with Hutu villagers who had participated in the butchering of the Tutsi in Rwanda. They described with some bewilderment how they had joyfully competed among themselves as to who would hack with a machete a larger number of their next-door neighbors, including many women and children. I would then ask students to relate what they saw and heard to debates about the driving forces and motivations behind such genocidal campaigns, sometimes carried out by thousands of apparently "willing executioners."[3] Or, in International Political Economy, I would show a segment describing a cashew-processing plant built with a loan from the World Bank in Tanzania which lay idle. I would then ask students about possible explanations for this glaring failure of international economic assistance, linking these to theories of economic development.

In addition to these pedagogical enhancements, I gradually pared down the PowerPoint outlines I projected in class and adopted a rule that I would never speak uninterrupted for more than a few minutes. I experimented with brief writing exercises, small-group discussions in class, and more practical group assignments outside of class. I worked hard to make my classes a lot more engaging for students, if not uninterrupted fun. As I was doing all this, I honestly believed I was becoming a better teacher and helping my students become better learners. According to the reigning pedagogical doctrine, those methodological improvements should have delivered some handsome intellectual returns (or superior "learning outcomes"). Yet, I observed with growing alarm an opposite trend.

I first became acutely aware that things were going amiss one afternoon in October 2002. That day Lejla, an obviously bright junior from Albania who was taking my Conflict and Conflict Resolution class, came to see me during my office hours. She was clutching a copy of "The Last Negotiation," a 4,250-word article from *Foreign Affairs* proposing a bold stab at a final settlement to the Israeli-Palestinian conflict.[4] Before presenting their own recipe, the two authors introduced briefly two alternative approaches to resolving that seemingly intractable conflict: one advocating incremental steps that would allow the two sides to gradually build mutual trust, and another envisioning an all-out effort to hammer out a comprehensive final settlement. Lejla told me with palpable frustration that the previous night she and a few other students had spent almost two hours arguing over which approach the authors favored, if any.

I was taken aback. To me, the position of the authors seemed crystal clear. After all, the title of the article itself reflected quite transparently the overall thrust of the argument it contained. Yet here were a group of intelligent, earnest students making an honest effort to understand that fairly uncomplicated argument—and falling short.

A few days later another student from the same class, Jovana, stopped by to share a different yet related kind of frustration. She came from Serbia and was one of the stronger students in the class. Jovana had no problem understanding each of the assigned articles or book chapters. She was concerned, though, that there were too many of those, and so much content could hardly "sink in." I tried to convince her that, by American standards, two articles or book chapters, on average 30-40 pages per class session (for a class which meets twice a week), was not an inordinate amount of reading. And once she graduated, she would need to process much larger piles of information in the field she would pick for her career. She left apparently unconvinced.

A few weeks into the following semester, in February 2003, two students in my International Political Economy class came in with different concerns. Sam, an exchange student from the United States, complained he could not really understand why we were reading most of the texts I had assigned and we were discussing in class. What purpose could all this theorizing serve? I could sense that the usual platitudes regarding the value of liberal education and critical thinking did not seem to convince him. Then Stanislav, an articulate third-year student from Bulgaria, asked why we were reading only Marxist authors. I took out the syllabus and tried to explain that he had the wrong impression: of the three texts with Marxist influences or references we had read, one author was, indeed, a quasi-Marxist; of the other two, one presented briefly some Marxist arguments against globalization, and the other was seeking to debunk such arguments. Stanislav also seemed unconvinced.

As I became more alert to the learning difficulties experienced by many of my students, I couldn't but notice that those were slowly but surely growing. With each passing semester, larger numbers of students seemed to struggle a bit more when asked to relate recent political events or debates to some of the main points in assigned readings or to broader concepts. They also found it harder to identify and compare the theoretical perspectives of different authors.

Following a few general themes running through a course (like the changing understandings of liberty in modern political thought, or the evolution of prescriptions for economic development) was becoming an ever greater challenge. Applying or just accessing most of the knowledge students should have acquired in previous courses was turning into a similarly daunting task. Responses to exam questions grew steadily fuzzier, and fewer students seemed capable of mastering the intricacies of coherent argumentation, logical transitions, paragraph structure, and English grammar and syntax.

The eroding command of written English, including vocabulary and spelling, seemed particularly unsettling. Even some of the best students, after four years of reading thousands and writing hundreds of pages in English, would still submit papers peppered with grammatical glitches or compose sentences with confused word order—reflecting the syntactic structure of their native language. They were also frequently failing to make some finer distinctions—for example between the different meanings of "economic" and "economical," "realist" and "realistic," etc. Many would refer to supporters of liberal ideas as "liberalists" (without recognizing the derogatory overtones of this label), and this was hardly the only term which had not become firmly etched into their brains.

Such problems were becoming particularly evident in the essays students wrote for their graduation exams. Those often included 400-word "paragraphs" devoid of much logic or coherence.[5] Research papers, including senior theses, seemed to demonstrate particular problems of organization and balance between developing a confident train of thought and bringing in appropriate references. Curiously, even some of the native speakers enrolled at AUBG as exchange students were showing some of these difficulties. A colleague once observed that some of the paragraphs they wrote looked as if they had tossed all the words they intended to use in the air—and left them where they had fallen on the page.

The Case for Cannibalism

This is the title of a brief, sarcastic article by Thedore Dalrymple, a conservative British social commentator.[6] It refers to an infamous case in Germany when someone posted an ad that he was looking for a person willing to be killed and eaten; a man responded; the two met, one killed the other, and consumed part of his flesh. The subtitle of Dalrymple's piece reads: "If everything is permissible between consenting adults, why not?" Throughout the text, he pokes fun at this liberal argument, seeking to demonstrate that it becomes absurd if taken to its logical conclusion. The last paragraph contains a telling observation: "The case is a *reductio ad absurdum* of the philosophy according to which individual desire is the only thing that counts in deciding what is permissible in society." Still, when I ask students in my "Introduction to Politics" class (some of them juniors and seniors) to work in groups and try to identify the ideological position of the author, they almost unfailingly conclude that he is a liberal. This is, no doubt, partly due to their inability to catch all the irony expressed in a foreign language. But over the last few years at least 15 native speakers (mostly exchange students) have participated in this exercise, and not one of them has been able to decipher Dalrymple's message.

Perhaps most disturbingly, at some point even discussions of current events started more often to fall flat as fewer students were regularly following the international news[7]—despite my frequent admonitions to do so and the inclusion of related bonus questions in mid-term exams. The first few semesters at AUBG I had skipped the introductory chapter of the international relations textbook I was using. Its main objective was to convince American undergraduates that international politics could, indeed, have a direct impact on their own lives, and was thus worth studying. At that point my students did not seem to need such a lesson.

In those days, students from Bulgaria and other Balkan countries still appeared to carry on a tradition started over a century earlier by their grandfathers. With a strong interest in international politics, they had engaged in heated discussions over the schemings of the "great powers," informed by newspaper articles read aloud in smoky village taverns. In those distant times, it had been immediately obvious that the fate of the squabbling Balkan tribes was to be largely decided by outside forces. That fatalistic outlook had engendered keen interest in the larger world and the political gambits played out in distant capitals.

The apparent interest of most students in political issues was additionally kindled by dramatic events like the ethnic wars raging in the former Yugoslavia. After the secession of Kosovo in 1999, AUBG accepted a few dozen students from the new entity, some of whom had personally experienced the conflict. Also, Bulgaria itself had recently gone through an acute political and economic crisis, and many students had vivid recollections from those events (some had even participated in protests against the "socialist," or reformed communist, government). Gradually, however, this kind of intense interest in politics seemed to fade. To most of my students, the larger social world was progressively becoming *terra incognita* much removed from their daily interests and concerns—a syndrome I knew had long become common among American students.

This is the bitter paradox I have faced as a highly committed teacher. I honestly believe my courses have become a lot more engaging and better designed. They have gradually incorporated more attractive (and still intellectually sophisticated and challenging) readings and multiple elements which should facilitate active student learning. I have developed a better understanding of my students as I have talked to many of them about their expectations and life experiences, and have read extensively about their generation's common attitudes. Also, I have become a lot more relaxed and confident in the classroom. Yet I have seen many of my students gradually grow more disengaged from the issues and ideas I would expect to be intrinsically interesting to any socially alert person. They have also increasingly struggled to master abstract concepts and the intricacies of written English. Consequently, they have started to complain more frequently that my classes are—that most damning of student put-downs—boring.

I have been unable to reverse this trend, no matter how hard I have tried. I have observed how the ability of the majority of my students to master complex issues, ideas, and English has continued to decline—luckily, with many notable exceptions. Yet the self-confidence of many seems to have mysteriously waxed, to the point where I have difficulties convincing some that their responses lack substance, coherence, or even sufficient grammatical precision.[8]

With Endless Gratitude

I cannot fully convey how indebted I am to some of the truly exceptional learners I have been lucky to teach.[9] I will limit myself here to three such examples. Aleksandra (from Serbia) and Klaudia (from Albania) both took my Introduction to Politics class, in the spring of 2010 and 2014 respectively. The course is designed around the ideas of a few modern political thinkers and contemporary ideologies, and the way these can be applied to contemporary political and social problems. Aleksanda and Klaudia both wrote after they had received their grades to thank me—the former saying how much she had enjoyed thinking through the questions on the final exam; the latter—that "there were few things that could beat the feeling that came from understanding seemingly inconsistent thoughts expressed by an author" (a sensation only strengthened by how much progress she had made in her ability to read through and grasp complex texts).

The most striking example, however, is provided by Yana, a healthily melancholic and very curious student from Turkmenistan who in the spring of 2014 asked me to supervise her senior thesis. At one point I commended her for the amount of relevant literature she had been able to read and successfully synthesize in writing her paper. She smiled and sighed: "Finally, I learned to read…" Still, her thesis suffered from a problem common to Russian speakers—since their native tongue has no definite and indefinite articles, they often fail to develop an intuitive sense of where these belong. Then, in September, I received a very long email from Yana (over 1,000 words). In it, she described her summer and asked for some advice related to courses she needed to take in her master's program at the Central European University in Budapest. To my surprise, the message had almost perfect grammar (with maybe just 2-3 missing or misplaced articles). I was most pleasantly surprised. It turned out Yana had again spent the summer working in the United States (on a work-and-travel program)—but this time she had read perhaps 30 books in English, mostly great novels. She said she had been inspired to read so voraciously by her work on the senior thesis—but even if this was a post-hoc rationalization, three months of keen reading had cured her lingering grammatical handicap. When students ask me how to improve their written English, I often refer to "the power of deep reading" (and I sometimes distribute a related handout in class). But I never expected such a rapid and dramatic effect.

While observing and mulling these troubling trends, I have resisted the temptation to blame and chide students for their learning difficulties as I have linked these to broader cultural and technological trends. These more general trends can explain why the cognitive anomalies I started observing over a decade ago among my students are not unique. I have heard similar complaints from colleagues teaching in the United States, Britain, the Netherlands, Bulgaria, and a few other countries.[10] Most of them have attributed their students' patchy knowledge and shaky academic skills to the fragmented nature of the typical college curriculum.

In my case, though, that variable had remained constant (and AUBG offers a more limited variety of courses as compared to a typical American college of similar size—1,000 or so students). Yet, the ability of the majority of my students to accumulate knowledge, develop their thinking, and master written English had somehow weakened. Neither my colleagues' theories nor the conventional pedagogical literature could help me untangle this conundrum. I therefore embarked on a quest to find a more credible solution to the educational mystery I confronted.

3

Neuroscience Rides to the Rescue

The start of my epic quest to solve the mystery of the educational micro-crisis I was observing coincided with a sharp increase in the number of neuroscientific studies using sophisticated brain imaging equipment. With much fanfare, the United States government had designated the 1990s the "decade of the brain" and allocated increased funds for brain research. The decade had lived up to the hype surprisingly well. It was marked by a true explosion of clever studies and experiments made possible by the introduction of smarter and more precise scanning technologies. The new computer-enhanced machines produced a flood of catchy images and fresh experimental results which quickly trickled down into the popular press. Some of the new findings seemed to have obvious relevance for the troubled field of education. To me, it also appeared intuitively obvious that a better understanding of the workings of the brain was essential for developing any new and improved methods of teaching and learning.

I did my best to read as many neuroscientific papers and articles as possible. Those often seemed, however, overloaded with technical jargon. Most were also too focused on finding clear-cut causal relationships between the processes ostensibly captured by brightly colored brain images and specific reactions of the human subjects stuck into the tunnels of scanning machines. So I continued to educate myself, relying to a greater extent on the guidance of neuroscientists who sought to draw some broader conclusions from their—and others'—experimental work. I also voraciously read books and articles written by neuroscientists and science writers (sometimes in tandem) who presented key findings in more digestible language and examined their larger social implications.

Among those multiplying "brain stories,"[1] there was one which immediately grabbed my attention. It was an interview with Martha Herbert, a child neurologist at Harvard specializing in autism research. The interview had been published in 2001 in an obscure independent periodical, *The Wild Duck Review*.[2] The whole issue purported to investigate the possible "End of Human Nature" in the face of recent technological advances which were transforming the natural and social habitat.

Some of Herbert's observations provoked uneasy thoughts in this respect. She warned that sensory overstimulation, a cocktail of potentially toxic chemicals,[3] ubiquitous electromagnetic fields,[4] many other environmental irritants, and "social/emotional derailments" were provoking unhealthy adaptations in the developing brains of American children. She feared that a neural Rubicon had been passed beyond which the rewiring of young brains was becoming irreversible. She was also worried that the "reductionist triumphalism" kindled by "techno-utopian visionaries" stood in the way of grasping the true significance of these existential threats.

Could the neural adaptations Herbert described be the clue I was looking for in my efforts to understand the cognitive changes I was observing in many of my students? I was intrigued, so I contacted Casey Walker, the environmental activist and intellectual who stood behind *The Wild Duck Review* and had conducted the interview. She recommended a book by Joseph Chilton Pearce, an aging child-rearing and education guru. The title of his book, *Evolution's End*, was reminiscent of the title Ms. Walker had chosen for the thematic issue she had produced.[5] It mixed insightful references to brain research with a whiff of New Age mysticism, and was full of prophesies which sounded darker than even Herbert's unsettling warnings.

Pearce's oracular pronouncements resonated with another disturbing book which attracted my attention around that time—

Jane Healy's *Endangered Minds*. Though published in 1990 and focused on the growing learning difficulties of American adolescents, it described with uncanny precision most of the symptoms I had observed in my students—from increasing knowledge and reasoning deficits, to muddled grammar and syntax. Like Pearce's account, it sought to relate cognitive tendencies to detrimental modifications in brain wiring. Healy's main claim was that these neural adaptations were being induced by an increasingly unhealthy social and technological environment. I wondered if these books could be depicting developments in which the United States had (as Tocqueville had expected) once again led the way, with other societies now following in their footsteps. That seemed a plausible conclusion. But contrary to my natural inclinations, I was not yet ready to succumb to the pessimism Healy's and Pearce's books so readily evoked.

My unwillingness to surrender to cultural and pedagogical gloom was bolstered by a much more hopeful book—*The Art of Changing the Brain* by James Zull.[6] Published in 2002, it was a bold harbinger of the whole flood of "neuroeducation" research and proselytizing which was to follow. Zull, a professor of biochemistry, drew on key findings from brain research to offer a better understanding of student learning. He described learning as a function of the proper activation of different brain areas, and their horizontal and vertical integration.

From Zull's point of view, the teacher's main task in any field of knowledge was the same—to help students achieve this optimal brain activation and integration. That goal could only be accomplished if students were actively involved in learning exercises and experiences which stimulated all their senses, triggered in them some excitement, and prompted them to make rich cross-references and associations—between newly acquired information and relevant background knowledge, between concrete facts or events and more abstract concepts, between personal observations or experiences and larger issues, etc.

For a time, Zull's overall argument seemed truly inspiring. I tried to follow faithfully his advice and think of yet more inventive ways to prod students to make the mental and neural connections he thought were so crucial. As already indicated, those efforts did not produce spectacular results. In fact, as I was trying to better understand my students' learning difficulties, I couldn't help but notice some troubling changes in my own mental functioning.

As a graduate student, in my late 20s and early 30s I had been rather proud of my intellectual sharpness. This mental self-confidence did not stem primarily from my outstanding performance in demanding classes or the quality of my research papers and dissertation. Rather, I compared myself favorably to most American graduate students and professors who often did something that struck me as profoundly odd. In a conversation or in class, they would tell an anecdote. Then, a few days later, they would tell the exact same anecdote, without the vaguest recollection of the previous conversation or class discussion. I observed those incidents with some amusement, and thought they reflected a peculiar cultural quirk. Until one day, after a few years of teaching, reading and writing, and juggling my other professional and personal responsibilities, I began doing the same thing.

Those troubling episodes started to occur with increasing frequency. They were later supplemented by an even more disturbing mnemonic hiccup. Once in a while, I would pick up an article and start reading it with interest. I would then turn the page and see with surprise a few highlighted passages or even brief comments in the margins. Those indicated clearly that I had read and mulled over the article a few months earlier. Yet, it had left no lasting trace in my brain. It seemed I faced an even greater problem than my students who could at least easily recall the lyrics of popular hits and the story lines of favorite movies. At first I thought that my worrying memory lapses were age-related. As I was pushing 40, perhaps my gray matter was beginning to atrophy, causing my memory to slip. But then, why had my fellow graduate students in the United States, all highly intelligent and knowledgeable, experienced similar symptoms in their 20s?

Survival of the Less Informed?

Curiously, many individuals with less intellectual occupations (and preoccupations) I was meeting at the time did not seem to suffer from the cognitive syndrome I was developing. In the spring of 2003, I took our family car to a garage in Blagoevgrad, the city of 70,000 inhabitants which hosts AUBG. The mechanic who greeted me, an earnest man in his 50s, looked at me and my Spartan Škoda, and asked matter-of-factly: "What happened to your Golf?" He was referring to our previous car, a Volkswagen, on which he had done some repairs two years earlier. Around the same time I went to a new hair salon close to our apartment in Sofia. The woman working there, also in her 50s, immediately recalled she had given me a haircut a few months earlier in another salon where she had worked at the time. Needless to say, I had no recollection of that encounter. I also had a few other similar experiences.

A book I picked up in 2005 at a used bookstore in London forced me to see the mental deficit I was experiencing in a new light. Written by David Shenk, it had a highly evocative title—*Data Smog*.[7] Published in 1997, which now seem like the good old days before Google, it described with much wit and concern the challenges of dealing with chronic information overload.[8] According to Shenk, opening the electronic floodgates with the advent of the internet was tremendously complicating the task of organizing relevant knowledge, storing what was worth keeping in long-term memory, and recalling it when needed.

Shenk argued that the only way to survive and thrive amidst this information glut was to adopt a strict, carefully considered "information diet." When food had become abundant and cheap in rich countries, steadfast control over its daily intake had turned into a vital necessity. The lack of such control led to obesity and an array of health risks associated with chronic overeating. In a similar way, smart people would now need to become highly selective about the quality and quantity of the information they sought to access, process, and commit to memory.

In my mind, I tried to combine these new insights with the lessons I had learned from Zull. It seemed that the judicious consumption of information Shenk advised required the kind of conceptual and neural mapping Zull thought teachers should foster in their students. There could be a problem, though. The density of the "data smog" itself could undercut the development of that conceptual and neural framework. The underdevelopment of such a framework could make it harder to see through the informational haze, assess what is significant, and allocate relevant facts and ideas to clearly marked mental folders. This vicious circle might explain many of the cognitive problems I was observing in my students. Still, I hoped that a more lucid awareness of these problems and redoubled efforts to address them, both on my part and in my students, could help cut through that cognitive knot.

Inspired, I poured that newly found wisdom into a brief article addressing the pitfalls of "teaching and learning in the age of audiovisual pollution."[9] Drawing on Zull's book, I advocated a relentless, reinvigorated effort to cajole students to make the necessary mental associations and neural connections. That increasingly hard task was to be accomplished through the use of lively readings, many vivid examples and various media, including the brief video clips I had started to show in class. I also presented this vision at a seminar for university faculty organized by the Bulgarian ministry of education. Later, I helped draft the section on liberal education in AUBG's five-year strategic plan. It restated the university's commitment to broader, holistic learning whose overall objective would be to assist students in developing a conceptual "grid"—a framework which would help them understand the larger social world and their place in it.[10] This vision also became part of the description of the political science major at AUBG.

As I was plotting new strategies for making the learning I and AUBG offered more meaningful and "sticky,"[11] I also tried to keep track of the exploding literature on "neuroeducation." Gradually, however, I started to lose faith in the efficacy of the prescriptions I found there, and in my newly acquired neuropedagogical wisdom.

4

The Limits of "Neuroeducation"

My skepticism regarding the promise of "neuroeducation" grew as I continued to observe in my teaching increasingly frequent cognitive incidents. Someone could say that perhaps my courses have held an attraction for weaker and apathetic students. As I already noted, however, many of the students who struggle to make sense of course readings and to refer to key points in them in exams (to say nothing of class discussions) seem earnest and fairly intelligent.

World Apart

The most telling examples of student confusion and detachment from the larger social world come from my Global Political Economy course. It is based on dozens of articles and book chapters. Each one of these is relatively brief and uncomplicated. Yet, in their totality at some point they started to overwhelm many of my students. In 2005-2006 I began to hear (and see in my teaching evaluations) increasing complaints that the amount of reading was excessive. I could myself observe that the scores of names (of authors, politicians, countries, international organizations, etc.), developments, theories, and viewpoints contained in the readings were inducing growing confusion.

The examples of such confusion from this one class are too many to list here. So I will cite just a few particularly striking ones, all coming from the spring of 2011—starting with my favorite. I included in the mid-term exam a bonus question asking students about the two terms commonly used to refer to the Chinese currency. We had discussed the controversy around the allegedly undervalued Chinese currency as part of the current-events roundups conducted at the start of each class. Its common name, yuan, had also been mentioned in at least two required course readings. Yet, out of 27 students, only five came up with the yuan. And of those five, two thought the other term for the Chinese currency was yen. Three had bothered to google the right answer after the class (which was, in fact, higher than usual).

The same open-book/notes exam also included a short-answer question requiring students to apply the logic of a counterintuitive concept, "comparative advantage," to a hypothetical situation. They had all taken macroeconomics, which is a course prerequisite; we had discussed the concept in class citing relevant examples; and I had distributed a related handout. The handout reproduced David Ricardo's famous two-countries/two-goods model and emphasized how his analysis was different from Adam Smith's, focusing on "comparative" as opposed to "absolute" advantage. Still, only three students were able to draw that distinction in a hypothetical situation.

Another question was related to a reading examining the work of the World Trade Organization. One highly conscientious sophomore had looked at the wrong reading and written about the World Bank as if it were the WTO. I later found out that after her first year she had been on the President's List, which meant she had earned a GPA above 3.80. Another very diligent student later submitted a briefing paper on an extraordinary meeting of the European Council called to discuss the unfolding crisis of the eurozone. The opening sentence referred to a "UN meeting in Brussels." Alas, these examples could go on and on.

I should add that the problems I see in my classes are quite similar to those reflected in political science "state exams."[1] While intended to be student-friendly, the format of the exam gives me and my colleagues some general sense of the intellectual sophistication with which political science majors leave AUBG. Students need to read in advance two articles which they can bring to the exam. There, they have two hours to write an essay addressing a question related to one or both of the assigned texts. While each semester we do get to read a few outstanding essays, most are far below what we dearly hope to see. In addition to the frequent writing problems, many essays do not even demonstrate a basic ability to understand the main points made by the authors—even when they write quite clearly for a general audience.

For the exam held in May 2011, we assigned two highly readable and provocative articles—one on the rise of "the new global elite" from *The Atlantic*,[2] and the other from *Foreign Affairs* describing the pitfalls of a "G-Zero world" (in which no major power, alliance, or international organization is really in charge).[3] The question students needed to address was related to a statement quoted by Chrystia Freeland, the journalist who had rubbed shoulders with many members of the "new global elite" she was writing about. She conveyed the opinion of a senior manager at a US-headquartered global company who thought "that if the transformation of the world economy lifts four people in China and India out of poverty and into the middle class, and meanwhile means one American drops out of the middle class, that's not such a bad trade." Students were asked to determine whether Ian Bremmer and Nouriel Roubini, the authors of the second article, would support this conclusion.

Many students, as they wrote, had problems drawing a distinction between the shared perspective of the "gloligarchs"[4] Freeland described and her own argument. Her straightforward criticism of the alleged detachment of the global newly rich from the communities hosting them had not seemed to help. Quite a few students also seemed unable to grasp the general point made by Bremmer and Roubini—that in a world devoid of effective leadership (or "hegemony") and enforcement, when national governments feel under pressure to address their citizens' concerns, the trade liberalization which had made possible vastly disproportionate rewards could hardly persist. Instead, it would likely give way to tensions and protectionist outbidding. Many students seemed to assume that Bremmer and Roubini were supporting the self-interested economic policies they said governments were under pressure to adopt. They were apparently confused by the modality Bremmer and Roubini used as they concluded that in the context of the global economic slowdown policy-makers everywhere "must worry first and foremost about growth and jobs at home."[5] Recognizing that Bremmer and Roubini, arguing from a more "realist" perspective, would generally be unlikely to share the American CEO's blasé attitude posed an even greater difficulty.

One way to interpret these problems is to see them as reflecting partly a lack of broad background knowledge and sufficiently rich associations. To illustrate the importance of such general knowledge, I always show in class a favorite one-minute video segment from YouTube. The clip comes from a press conference at which Dana Perino, President George W. Bush's then young press secretary, shrugs off an inconvenient question. A journalist asks her whether President Putin had made a valid analogy when he compared the prospective deployment of components of the United States' anti-ballistic missile shield in a few East European countries to the Soviet actions that provoked the Cuban missile crisis back in the early 1960s. Perino looks unfazed as she starts to recite platitudes not directly related to the question posed to her. I ask students whether they think she addressed the question effectively, and they usually laugh. I then ask them why Perino did not exactly shine as the chief White House communicator.

Students usually respond that it must come naturally to a political operative to spew meaningless banalities. Some even think this is a vital skill for any government bureaucrat or politician. Occasionally, someone would suggest that Perino was perhaps concealing some terrible secret. Very rarely does someone stumble upon the right answer—that the White House spokesperson was unable to recall what the Cuban missile crisis had been all about. And that had not been an understandable momentary blackout under the pressure of the moment, the flashing lights, and the dozens of cameras directed at her. Only after she got back home in the evening and asked her husband was she able to recall the main events of the crisis. She acknowledged as much on a subsequent NPR show, without betraying much embarrassment at her own ignorance. Like most of my students, she had lacked the urge to find the response online before she left work.

Students sometimes find it difficult to believe this story. Particularly when I tell them that Perino had received a bachelor's degree in mass communications with a minor in political science; and had later earned a master's in public affairs reporting. Even if the Cuban missile crisis had not been covered in any high school history class, she must have later taken some college courses making references to it. The crisis is often mentioned in articles and books as one of the main events of the Cold War, and its unfolding is also depicted in two Hollywood movies. How can someone with this background, who has taken up a political vocation, have no recollection of such a central event in recent American and world history?

The point I want to make with this striking example of high-level ignorance is not that insufficient knowledge could be an obstacle to a high-flying career. Perino obviously had one (and has flown ever higher after leaving the White House). And, as President George W. Bush famously joked, a "C" student could still become president of the United States. Rather, I want to demonstrate to students that key political events cannot be stored in long-term memory and be easily recalled if they remain disjointed bits and pieces of largely irrelevant "information" (a problem Republican vice-presidential candidate Sarah Palin had in an even more obvious way when she was interrogated back in 2008 about the "Bush doctrine" and other foreign affairs topics). Since human memory works by association, facts removed from immediate personal experiences must be related to a larger framework and an overall web of significance.

Many students seem receptive to this message. The problem, though, is that even if they have the desire to build such a frame of reference, many struggle to do so. Students are often caught in a catch 22 (though many would not know the precise meaning or origin of this phrase), or the vicious circle I mentioned above: in order to be able to relate to political events and larger issues and accumulate knowledge about them, they need to make rich associations with a larger conceptual framework and a reservoir of background knowledge; and to build this framework and acquire such background knowledge they must find political events and larger issues relevant, if not inherently interesting. This has become the hardest cognitive nut to crack, and a source of mounting student frustration or dejection.

Once in a while, a student would speak out and offer an explanation for Perino's alleged ignorance which many others probably share but are shy to articulate: she is a typical American; and, as we all know, Americans have precious little knowledge about anything. This conviction is reinforced by many popular clips on YouTube showing mostly young Americans baffled by silly questions. To test the hypothesis that non-American students must have superior knowledge on most larger issues, I occasionally gave students an improvised, anonymous general knowledge quiz. The results from one I administered in the fall of 2009 are quite typical. As it turned out, they were roughly comparable to those of American college students.[6]

Terra Incognita

The quiz I asked my students to take was intended to test their general knowledge of world history and geography. I distributed it in my introductory political science course, and it was completed by 58 students divided into two sections. Less than half were able to recognize the Soviet Union as a World War II ally of the United States in a multiple-choice question, though almost a quarter came from former Soviet republics. Seventeen and 32 percent respectively could identify Israel and Iraq on a political map of the larger Middle East (with no country names on it). Only a few students got both right, with some writing "Israel" over Morocco or "Iraq" over Nepal (or other similarly far removed countries). Of course there have been some heartening exceptions. For example, in the spring of 2014, again in my introductory course, I mentioned the Indian Mutiny, and gave the wrong year for it (1859 instead of 1857). Stefani, an extremely alert first-year student from Bulgaria, corrected me—and this was one of the happiest moments in my teaching career. Unfortunately, such incidents have been quite rare.[7]

In addition to difficulties identifying the main ideas in course readings, building a coherent conceptual framework, and associating concrete events and developments with it, students have also become increasingly confused about major and minor requirements, course prerequisites, registration procedures, etc. The reason for this overall confusion could be that AUBG has started to admit weaker students. Still, the average SAT scores of the close to 300 students who entered the university in fall 2010 were 1185 (from the Critical Reading and Math sections of the exam). This was a bit lower than a decade earlier[8] but was still a fairly respectable score, particularly for non-native speakers.

Moreover, I have observed rapid changes in the thinking and sensibilities of even the brightest students AUBG still, luckily, attracts. First and foremost, their outlook has become a lot more utilitarian. As a result, they have largely lost interest in academic careers and Ph.D. programs. Most shun even the typical two-year American master's programs in favor of one-year, generally less demanding West European ones—which still offer the credentials they seek in order to receive a head start in whatever career they plan to pursue. Also, even the best students I see now seem a lot less animated by debates around larger ideas. Many are easily turned off by readings they deem too "theoretical," and I am not referring to French-style intellectual obscurantism.

Most strikingly, a decade ago the majority of students writing a senior thesis took a rather sudden intellectual turn. As we have lost some of our confidence in the ability of most students to produce a major research paper, we have made this a rather elitist exercise at our department. Perhaps a tenth of the political science and European studies majors graduating each year wish and qualify to write a thesis. In the past, they would typically produce narrative texts that address broader questions like the causes of the rise of ethnic nationalism and the dissolution of Yugoslavia—and sought to draw more general conclusions. Then, in the spring of 2006, almost all picked very narrow research questions which could be tackled with various statistical techniques or other empirical tools.

This shift came quite abruptly, without direct prodding from anyone. Students apparently thought that was the only way they could give their research papers substance and validity, no matter how questionable their underlying assumptions or unrepresentative their samples. They also appeared to feel more comfortable looking at specific, fairly concrete issues largely detached from a broader social or intellectual context. Over the last 2-3 years, the fascination with statistical techniques has subsided a bit, but not the search for a narrow empirical focus.

The End of Theorizing?

Sometimes reliance on statistical validity produces really striking results. For example, in April 2011 a student defended a senior thesis exploring Bulgarian attitudes toward Turkey and Bulgaria's Turkish minority. It was a quantitative study based on an obviously unrepresentative sample. Moreover, some of the interpretations it offered seemed driven by a clear ideological agenda (to prove the lack of strong biases and hostility). The student was one of the brightest and had one of the highest GPAs among all graduating majors. In an effort to alert her to some of the limitations of this line of research, I sent her an article by Jonah Lehrer on "the decline effect."[9] It describes the difficulty medical researchers and psychologists have had trying to "replicate" controlled experiments in order to confirm their validity. The article concludes with the following observation: "When the experiments are done, we still have to choose what to believe." The student wrote back to thank me, but she had drawn a very peculiar lesson—she realized she should not try to repeat her study and confirm its findings in graduate school, since she was unlikely to obtain the same results.

As I was ruminating about these inauspicious trends, I had one consolation. I knew I was not alone in this situation. In addition to the complaints of colleagues in different countries I mentioned earlier, I frequently recalled a 1997 article by English professor Mark Edmundson.[10] In it, he lamented that most of his students could no longer relate to the tragic worldview offered by, say, Sigmund Freud, and were consequently judging his courses and teaching mostly on the basis of their entertainment value.

Among more recent musings on this issue, one book clearly stood out as a similar confirmation of my deepest anxieties. In it, another English professor, Mark Bauerlein, proclaimed the current crop of American college students "the dumbest generation."[11] That title was, no doubt, a calculated provocation and marketing ploy. It also seems a bit unfair since the kind of ignorance Bauerlein uncovers is not limited to the Millennial generation.[12] But most of the statistics cited in the book, even if a bit selective, seemed shocking in their own right. At that point, I had already lost hope that the technical solutions offered by the exploding literature on "neuroeducation" could help address a problem which seemed so profound and widespread.

5

The End of Authority

I am not sure why it took me so long to become so skeptical regarding the bright future neuroscience could plot for education. With hindsight, this is particularly puzzling since my observations of the growing learning difficulties experienced by my students were supplemented by other worrying signs that should have given me pause.

After I became a parent and a full-time college teacher in the late 1990s, I started to pay more attention to the usual complaints that something in the way children and adolescents were growing up was terribly amiss. I was reading or hearing about numerous incidents indicating that this time around maybe the kids weren't alright. I also witnessed a few episodes when children and adolescents seemed to behave more badly than the kids I had played with back in the 1970s. But those concerns and impressions somehow did not strike a sufficiently strong chord—until one encounter with a group of Bulgarian teenagers which shook me out of this complacency.

Still in elementary school, our daughter Gali was developing a mild obsession with Rihanna, as well as some of the knee-jerk skepticism typical of our native country (and nuclear family). As a result, she had no hope of seeing Rihanna in concert in Sofia (Bulgaria's capital) before they had both grown up. Gali thus went truly ecstatic when in the fall of 2007 she heard that Rihanna was to perform at a free concert in the city's central square, courtesy of a cell phone company vying for the hearts of young users. How, then, could we refuse when Gali begged us to take her to the concert?

We arrived an hour early in order to secure a better observation spot. As we were approaching the venue, we passed by groups of 14-15-year-old girls who were milling around and chatting. Many had bottles of beer in their hands, and some were also smoking. In Bulgaria, teenage rebellion has made such casual, public consumption of alcohol and nicotine by teenagers the new normal.[1] So we tried to laugh it off with a few sarcastic remarks. What followed, though, was a lot more difficult to take in stride.

In the middle of the crowd, there was a minivan unwisely left by its owners in the square (which is normally used as a public parking lot). A group of girls and a few boys who wanted to get a better view of the stage quickly climbed on top of the vehicle. As they heated up to the loud music, they started to bounce up and down upon that improvised dancing platform. A few adults tried sternly to cool them down, or even urged them to climb down from the gyrating minivan.

The teenagers cheerfully rose to that challenge. They responded to the criticisms directed at them with utter self-confidence, composure, and a sense of humor. They taunted and heckled the adults, occasionally entering into mock arguments with them. They also made a few vague threats. The adolescents laughed back at warnings that the police would be called in. They did not for a moment consider surrendering their elevated position.

What seemed most unsettling in the behavior of the whole gang was not any deliberate effort to challenge adult authority. Rather, it was the complete absence of any palpable awareness that such authority could exist, even hypothetically. The teenagers did not seem to sense that since the people they were addressing were 20 or 30 years their senior, they might deserve some basic respect and courtesy. They talked back and hurled mock insults in the way I imagined they would respond to classmates who teased or challenged them in the school yard.

During the family debriefing we conducted after the concert, Gali reminded me of stories she had told me about her classmates since kindergarten. There was the six-year old boy who had angrily berated his teacher: "Don't look at me like that!" And when she asked a bit surprised, "Are you talking to me?" he retorted, "Yes, if you aren't deaf!" And the eight-year-old boy who was brought to the Stalinist school principal to be disciplined for punching another boy hard in the stomach; and who instead lectured her self-assuredly about his own emotional needs. And the classmates who could not sit still in class, so would get up from their seats, talk loudly and laugh, and ignore the half-hearted attempts of the teacher to get them to observe the most basic rules. All this should have convinced me that the behavior I had observed was quite unexceptional. Still, the scene seemed inordinately distressing.

A few weeks later, I witnessed a similar incident. Six teenagers (four boys and two girls, maybe 15 years old) were riding on a Sofia tram. They were talking and laughing quite loudly. At some point, one of the boys dropped a large plastic bottle of beer, which was left to roll around on the floor for a while. Two elderly women tried to scold the youngsters mildly before getting off the tram. Their remarks were met with casual derision. Then two of the boys started boasting how they would never validate a ticket on public transport, and would never ever pay the fine if caught. The rest laughed with knowing approval.

At that point, the group was approached by a ticket inspector. It turned out that four of the teenagers had, after all, valid passes provided by their less rebellious parents. The two boys who prided themselves on always traveling gratis did take crumpled tickets out of their pockets. Laughing loudly, they punched those before the eyes of the resigned official, and requested with mock respect to have them checked. One boy took his cell phone to take a picture of—allegedly—the only ticket he had punched in his life. After the inspector walked away shaking his head, they continued to laugh and shout, with one of the girls sitting on the lap of one of the boys (without any sign that they were romantically involved).

A few days later, my anxieties were reinforced further by an exercise of deliberative democracy on Bulgaria's state-owned TV channel. It aired a discussion related to the plans of the Bulgarian ministry of education to introduce a new set of national tests for graduating high school students. The main participants were the minister of education (a self-assured lawyer in his late 40s), a deputy-rector of Sofia University (an academic in his 50s), a prominent Bulgarian sociologist-turned-pollster-turned-pundit (probably close to 60 at the time), and a group of five high-school students.

The group of students included four young women and only one young man whose grotesque behavior seemed truly out of place. He was constantly interrupting the adults with derogatory remarks. At one point he started waving an improvised cartoon illustrating the problematic pedagogical doctrine and practices promoted by the ministry of education. He then stood up and took the cartoon to the government minister, lambasting him for his ostensible failure to give satisfactory answers to the students' questions.

Throughout the debate, the young man was behaving like a clown, full of—and utterly pleased with—himself. He might have been right regarding some of the deeper problems besetting Bulgarian education. Incidentally, his own cognitive difficulties were a good illustration of those, for he proved clearly incapable of following and addressing the arguments directed at him and the other students. But his inability to sustain coherent conversation was obscured by his complete failure to recognize that the situation required him to show some respect, or at least minimal politeness.

Following those incidents, I wrote an anguished op-ed piece for a Bulgarian paper linking the crisis of secondary education in Bulgaria (confirmed by many international studies[2]) to the collapse of propriety and respect for adult authority I had witnessed. Still, I did not quite see the implications of the attitude and behavioral problems I described for college education. It took another two years or so for a new flash of insight to really change my perspective.

Part of the reason why that did not happen faster is that most of my friends and intellectually significant others did not share my concerns regarding the examples of gratuitous misbehavior I had observed. Most of them thought I was overreacting to the understandable—and perennial—teenage penchant to push the boundaries and challenge adult authority. I, somehow, could not embrace the idea that the scenes I had witnessed were just innocent illustrations of what had always been transitory, maybe even necessary stages in the healthy development of youngsters seeking to push back against overbearing adult authority—maybe because as a teenager I had never felt a need to do this in the stereotypical way.

The sympathetic view of youth rebellion I encountered is, in fact, quite common. It is often supported with references to a few quotes which are splattered all over the Internet:

> I see no hope for the future of our people if they are dependent on the frivolous youth of today, for certainly all youth are reckless beyond words. When I was a boy, we were taught to be discreet and respectful of elders, but the present youth are exceedingly wise [clever] and impatient of restraint.

> The children now love luxury; they have bad manners, contempt for authority; they show disrespect for elders and love chatter in place of exercise. Children are now tyrants, not the servants of their households. They no longer rise when elders enter the room. They contradict their parents, chatter before company, gobble up dainties at the table, cross their legs, and tyrannize their teachers.

> The world is passing through troublous times. The young people of today think of nothing but themselves. They have no reverence for parents or old age. They are impatient of all restraint. They talk as if they knew everything, and what passes for wisdom with us is foolishness with them. As for the girls, they are forward, immodest and unladylike in speech, behavior and dress.

These laments seem to raise some fairly recent concerns. But though not securely dated, they apparently all have a distant origin.

The first quote ostensibly comes from Hesiod, a Greek oral poet who probably lived in the 8th century BC. The second one is often attributed to Socrates, but its source is similarly uncertain. The third is said to originate from a sermon delivered by Peter the Hermit, a fiery French priest whose eventful life ended in the early 12th century AD. While these attributions are questionable, the quotes are used to illustrate that worries about the young are probably older than the mini-skirt, the hippy outfit, the punk haircut, the multiple piercings and tattoos, and the raucous behaviors associated with spoiled youth since the 1960s. The idea is that complaints about the young generation have always been around, and the kids have still turned out all right—as is likely to be the case now.

Before the youth rebellion of the 1960s, some earlier social critics had already greeted with apprehension the cultural and generational shifts accompanying the wave of industrialization, the rapid expansion of the market economy, and the accompanying spread of liberal ideas and attitudes in the 19th century. Such apprehension, even stronger and more pointed than the ancient and medieval examples cited above, was evident in the writings of prominent British cultural conservatives like Samuel T. Coleridge, Thomas Carlyle, John Ruskin, and Gilbert K. Chesterton. At that time, there were particular fears that boys and young men were becoming weak and effeminate. As a counter-measure, in the late 19th century Baron Pierre de Coubertin launched the modern Olympics. Those were billed as a reincarnation of the ancient Greek games, and de Coubertin was particularly inspired by the love of sports he had observed in Britain. But apparently boys there were in trouble, too. So a few years later British general Robert Baden-Powell founded the Boy Scout movement, which quickly spread to the United States and other countries.

By all accounts, those earlier grumbles had proven a bit overblown. Could, then, current concerns about the young generation be as misplaced? This is the view taken by most liberal academics. They habitually describe excessive worries about youth misbehavior as instances of unwarranted "moral panics," whipped up by conservatives clinging to an outdated model of oppressive authority. Such fear-mongering is dismissed as a conspiracy sometimes assisted—knowingly or unknowingly—by other interest groups and the sensationalizing media.[3]

Consider, for example, the birth of a British baby in early 2009. The father was initially thought to be a 13-year-old baby-faced boy who stood four feet (122 cm) tall and appeared even younger than his age. He had been only 12 at the time of conception, and the mother was a 15-year old girl who looked in photos as if she could be his mother. The young dad's sister had herself given birth at 13, and his father had left the family a couple of years earlier for a teenage girlfriend. Both young parents came from families living on welfare, and the boy was getting a small allowance. He had only the vaguest grasp of his looming parental responsibilities, yet announced he was ready to assume these. The pictures of the new "family" were splashed all over Britain's infamous tabloids. It later turned out that the real father of the baby was another, 14-year-old, boy.[4] Still, the whole affair provoked the predictable howls from social conservatives about the plight of "broken Britain."

This ominously alliterated catchphrase seemed to capture a host of rising concerns regarding objectionable behaviors by children and teenagers in Britain, as well as in other countries. Recent years have seen a flurry of publications raising worries about all kinds of related problems: from increased impulsivity and distractibility, to aggressiveness, "anti-social behavior," self-injury and a stubborn unwillingness to grow up.[5] Such stories do reflect wider public concerns. A few years ago an opinion poll in the United States found that "relatively few parents believe they have been successful in teaching their children many of the values they consider 'absolutely essential'." Most respondents "describe[d] teens and children with words like 'lazy' and 'irresponsible'," and few of them agreed it was "very common to find young people who are friendly and respectful." About a third agreed that the younger generation were "spoiled."[6]

Most liberal academics and intellectuals have one common response to all similar concerns. Hand-wringing about youth crime and anti-social behavior? Apprehensions regarding the violence, profanity, or sexualization permeating TV shows, music video clips, video games, and pop culture in general? Worries about the rising numbers of disorganized and apathetic students, mostly boys, who have lost all motivation for learning and even social success? Worries about vicious bullying and physical aggression, girls fighting like boys, sexually active adolescents, teenage pregnancy, and sexually-transmitted diseases? Concerns about an "obesity pandemic"? Any other alarmist howls? They all reflect attempts to whip "moral panics" in response to innocent or even liberating challenges to social prohibitions, with the thinly veiled intention of keeping oppressed groups—and children—down. Such panics are said to be fanned despite much scientific data establishing that, in fact, youth crime, "anti-social behavior," pregnancy rates, etc. have declined, and teenagers have in many respects become more conservative than their own parents had been. And exceptions like the young British parents (who had perhaps not received proper sexual education) can only serve to prove the rule.

This upbeat approach is aptly illustrated by the work of two Australian sociologists who designed a clever "moral panic neutralization project."[7] They offered their support to an "outlaw motorcycle club" whose members' allegedly rowdy behavior had been met by a predictable campaign to drum up a "moral panic" in the local media. Multiple news stories and opinion pieces had identified the bikers as "folk devils," scapegoats onto whom journalists and the public could project broader social anxieties. The authors went ahead to form a coalition with the alleged "folk devils." As part of an integrated "action research process," they launched a vigorous counter-scapegoating media campaign. The brave fight they put up ostensibly helped produce a dramatic reversal of attitudes in editorials and among the concerned public. According to the authors, this happy ending provided "an example of a macro-level intervention through which liberation from oppression was affected."

This commitment to the liberation of oppressed groups is commonly shared by liberal intellectuals and academics. The tendency is well illustrated by a study of the "moral panic" around the alleged spread of violent and aggressive behavior among Canadian teenage girls in the late 1990s.[8] That panic was stirred by a much discussed TV documentary and an avalanche of news stories decrying the exploits of "nasty girls." These accounts appeared in the aftermath of the highly publicized killing in 1997 of a teenage girl by a group of classmates, mostly girls. They had accused the victim of gossiping about one of them, talking to the boyfriend of another, and snooping into the address book of a third.

Obsession with this and a few other horror stories had spread—despite reliable evidence that the number of girls charged for murder and attempted murder had remained at a constant low since the 1970s, and youth violent crime in general had declined slightly in Canada. According to the authors' diagnosis, "the moral panic over the Nasty Girl is part of a backlash against feminism."[9] In their view, such overblown alarmism "over the statistically insignificant Nasty Girl is a projection of a desire to retrieve a patriarchal social order characterized by gender conformity."[10] Curiously, such "liberal" arguments are endorsed by right-wing libertarians, like those writing for *Reason* or the anonymous authors of *The Economist*.[11] They typically have a hard time mustering much moral outrage, and worry that concerns over youth misbehavior can be used to justify unwarranted government overreach across many areas.

Don't Worry, Be Happy

Among commentators in the media, Virginia Heffernan seems to have the least patience for any alarmist, morally-tinged hand-wringing. A case in point is a review she wrote for an HBO documentary series, *Addiction*, which in 2007 sought to dramatize the effects of various kinds of substance abuse in the United States. Heffernan ridiculed the morbid fascination of Americans with alleged addiction epidemics, fed in this case by an authoritarian TV channel bent on sensationalist fear-mongering. On the surface, her sarcasm was directed against the overall message of the documentary that "drug and alcohol addiction are diseases of the brain, and they can be treated, at least partly, with medicine." But the way Heffernan satirized an ostensible public "addiction" to addiction documentaries implies that she has a more general problem with the dire warnings they contain:

Pseudoscientists don't know yet whether drug-documentary addicts are hooked by the gruesomely

> thrilling scenes of tourniquets and needles, the photos of pre-Vicodin fifth graders or the promise of redemption through higher powers. But something definitely sets the brain reeling with manic questions: How could they fall so far? How could so many of us? Whom will addiction strike next, and will the culprit be the demon rum or the demon OxyContin? ... The blunt title holds promise. As a story, addiction to drugs and alcohol has a chilling and ritualistic arc. Typically, the variable is the drug. Some viewers go for the methamphetamine documentaries, with their slightly high-handed attitude toward the Midwest, their contested statistics and their focus on dental issues. Other viewers prefer the shadowy, stylish heroin ones, with the sexy, skinny children and "Requiem for a Dream" fashion.[12]

These arguments may seem quite logical. But, as I already indicated, there is much brain research indicating that our embrace of ideas, values, political platforms, and candidates is not a matter of pure logic. Instead, it is driven largely by automatic affective reactions, or "gut feelings"—or a (near) lack of such neurosomatic responses.[13] Musing about my own intellectual evolution, I have come to see myself as a prime example of this tendency. The night of the Rihanna concert, something broke in me; and that sinking feeling has not left me ever since. I knew very well that the behavior I had observed was quite innocuous compared, for example, to the varieties of "anti-social behavior" often reported in the British press (sometimes involving unprovoked, vicious attacks on adults by adolescents). Still, there was something that struck me as profoundly odd in the behavior of the Bulgarian teenagers I had observed—their total lack of awareness of any vertical social distance between themselves and the adults they were confronting with casual nonchalance.

Following that incident, I found it increasingly difficult to shrug off the defiant youth language and behavior I observed, heard of, or read about as representing an unexceptional, passing stage in a process of overall personal maturation. And I did drift toward the conclusion that 1) the adolescent attitudes and behavior I started to find so disturbing, and 2) the cognitive problems I was observing among my students, were somehow related. In making this connection, I did not have in mind the usual understanding that some students with learning or behavioral problems could have more general difficulties sustaining their attention and controlling erratic impulses. Rather, I had a growing sense that there was something a lot deeper connecting these two tendencies. For a time, I could not quite put a conceptual finger on it. Until it all clicked together in a flash of insight one evening in May 2011, as I was unwinding from grading final exams.

6

My Stroke of Insight

My epiphany came, as usually happens, quite unexpectedly, after I had serendipitously read in rapid succession five quite different articles.[1] The first of these described the extent to which the formation of strong memories depends on the emotional arousal of learners as they digest information or experience various events.[2] The second one addressed the anxieties of parents of children whose extraordinary emotional intensity made them psychologically vulnerable, while also immensely curious and capable of acquiring rich knowledge on the go.[3] The third explained how cocaine can desensitize the brains of addicts and thus rob them of any other sources of excitement—to a point where even that high no longer feels gratifying.[4] The fourth was a heart-wrenching personal account by a self-described gambling addict indicating that unhealthy compulsions which did not involve substance abuse could have very similar effects.[5] And the fifth was Nicholas Carr's now famous provocation, "Is Google Making Us Stupid?", which I reread before I opened the book he had spun out of it.[6] Back in 2008, he had already argued that excessive internet use could induce changes in brain wiring—making sustained attention and the pleasures of "deep reading" elusive, and even scrambling our emotional responses and thinking.

Those five articles were only loosely related, yet the main impressions they left in my mind somehow clicked as the pieces of a mental puzzle. Here is how I instantaneously connected the dots. If long-term memorization depends on affective response, could many of my students have trouble storing in memory and recalling key facts and ideas because those were not sufficiently arousing? In other words, the sensibilities of students were perhaps moving in the opposite direction, compared to those children and young people living with extraordinary emotional excitability. Was this partial desensitization and detachment akin to the effects of cocaine and gambling on the brain—but induced instead by all those hours of browsing, clicking, scrolling, watching, shooting, scoring, and interacting on (or with) the internet? And could this addiction to digital stimulation, like all addictions, partly shut down the prefrontal cortex (which, in any case, is not adequately developed until around the age of 25)—creating problems with sustained attention, reasoning, and appropriate inhibition? Were students experiencing this syndrome losing simultaneously some of their emotional attunement, curiosity and mental sharpness—as well as their sense of propriety, including their awareness of adult authority and the necessity of key social norms?

As eureka moments go, that flash of insight produced a powerful surge of emotion in my own heart and brain. Needless to say, I interpreted that as proof that I had struck conceptual gold. Over the following week, I briefed a few friends on the bold—and depressing—learning theory which had taken shape in my mind. As I did this, I became increasingly convinced in the correctness of my educational diagnosis. I continued to read avariciously on the emotional basis of learning and memory, the biological mechanism of addiction, the effects of internet use on the surfing brain, learning broadly conceived, and related issues. As I did this, I naturally picked up observations and conclusions which reinforced my own inferences. I also added a few insights related to implicit learning and what I saw as growing existential disconnect.

It took me another two years or so to flesh my skeletal conjecture—and to acquire sufficient confidence to present it in two rather amorphous conference papers. Those were both quite pessimistic in tone—and then it got worse. I gradually came to see the learning and behavioral problems I was trying to demystify as part of an even larger psychocultural syndrome. Later, I even developed new apprehensions about the success of some students in developing the "analytical skills" routinely included among desirable "learning outcomes" in college syllabi and educational or promotional materials.

7

A Well-Tempered Brain[1]

I wish I could use "addiction" and its related words in a loosely metaphorical sense, as a way of dramatizing a problem that should look gravely serious in itself. I am afraid, though, that the syndrome I want to describe deserves the label literally—whether it meets the formal criteria for a distinct mental "disorder" handed down by the American Psychiatric Association or not.

If I had any doubts in this respect, they were dispelled in July 2011 when I attended a high-profile international conference on "teaching democracy" in Oslo, Norway. It was an eye-opening experience, for reasons slightly different from those intended by the organizers. On the bus from the airport, a 14-15-year-old boy in the seat in front spent maybe 40 minutes on Facebook—scrolling up and down on his smartphone, typing comments, smiling occasionally, and doing whatever you typically do on Facebook. Once in a while, he would put down the device and try to look out of the window, but in a minute or two would feel an itch to resume his scrolling and ritual touch-screening. It was a vivid reminder that the information revolution was not necessarily enhancing interest in, and perusal of, information related to the larger social world. During the conference, however, I was amazed to see that boy's behavior replicated on a much grander scale.

Almost half of those present (some of them college or university presidents and provosts, many in their 50s and 60s) spent much of the time tapping on various electronic devices, mostly tablets connected wirelessly to the internet. They did that without any palpable sense of embarrassment, as if it was the most normal thing to do in such settings. During one plenary session, a woman in her late 30s snuck in late, settled in the seat next to me, and pulled out a tablet. She then spent almost an hour toggling frequently between Facebook, Gmail, Jezabel, a few shopping websites, and whatnot. Like the boy on the bus, she looked up from the screen a couple of times and tried to tune in to the speaker in front, but that never lasted long. I thought such behavior could serve as an ironic paradigm for understanding one of the main reasons for the apparent flagging of civic education at a moment when, as speaker after speaker emphasized, it was most needed.[2]

If it was so difficult for many of these educational leaders to look away from the screens in front of them, I wondered what a more geeky gathering would look like. I did not have to wonder long, for I stumbled upon Nicholas Carr's vivid account of one such event. He had attended a conference on information technology during which most participants never really put down their smartphones and tablets. Whether they were attending a presentation or speaking to someone, they appeared to treat the words coming from the mouth of the flesh-and-blood speaker before them as just one data channel among many—to which they could not be reasonably expected to allocate their undivided attention.[3]

Of course, deploying this kind of "continuous partial attention"[4] could be seen as a deliberate strategy—a calculated, often enthusiastic embrace of the level of multi-tasking we will all one day need in order to function productively in an IT-saturated milieu.[5] But I strongly doubt this is the whole story, and I am not alone. The notion that surfing, clicking, scrolling, app manipulation, instant messaging, gaming, and other digital activities or touch-screen routines can become addictive has already received support from some psychologists, psychiatrists, and neuroscientists.[6]

This idea has also spread in the popular press and imagination since it resonates with some strong worries among parents and teachers in the United States and other wealthier countries.[7] They have increasingly complained that they find it difficult to get many children to switch their attention to less compelling activities or forms of communication traditionally associated with schooling and even play. Concerns about some kind of digital addiction also reflect the anxieties of skeptics who feel, like Carr, that they have lost as grown-ups the mental quiet and sustained concentration needed for "deep reading."[8]

Worries about compulsive digital interaction, especially online video-gaming, are particularly strong in some Asian countries where such habits have already produced visible health problems and other consequences, and have been accepted as meriting a clinical diagnosis. These have been most dramatically illustrated by a few tragic incidents. For example, in recent years several young men have died after prolonged gaming sessions. Those sometimes lasted two or three days, and during that time the victims did not sleep, barely ate, and eventually suffered physiological shut-down as a result of utter exhaustion.[9]

In one particularly shocking—yet telling—case, a South Korean couple left their three-month-old prematurely born daughter to starve to death in their apartment. While she was fading away, they spent long stretches of time at an internet café. There, they were engrossed in a popular role-playing online game in which their task was to nurture a virtual baby girl.[10] Even in the absence of such lethal outcomes, the medical authorities in South Korea, China, and a few other countries have recognized internet addiction as a mental disorder and a major public health problem. They have consequently opened specialized clinics, sometimes employing very harsh methods, as part of efforts to curb perceived epidemics of compulsive gaming and other sorts of digital overindulgence.[11]

Some American psychologists are similarly apprehensive. Kimberly Young was the first to suggest a similar diagnosis.[12] Over two decades ago, she founded the Center for Internet Addiction offering treatment (a lot less harsh than the Chinese variety) for the condition. Victoria Dunckley has cast an even wider net. She has attributed the tendency of many children and adolescents to act out and be easily distracted (to the point where they are diagnosed with ADHD or other disorders) to the hyper-arousal triggered by chronic exposure to digital screens. In her view, interactive features reinforce the addictive potential that should have been obvious in the case of TV.[13] In her blog dubbed *Mental Wealth* (hosted by *Psychology Today*) she makes tireless efforts to alert parents and educators about the increasing danger posed by the saturation of daily life with digital technology. Both psychologists offer recovery programs involving digital abstinence—with Young recommending a "digital detox" and Dunckley advising an "electronic fast."[14]

Why is the unrelenting siren call of the internet and digital screens so irresistible for so many—particularly, but not only, the young? This question cannot be addressed without a vague comprehension of how the human brain processes information coming through the senses and from the body, and generates mental and behavioral responses to these "inputs." The most relevant fact about the brain is that throughout its evolution it has acquired an elaborately layered organization. Parts which have developed more recently in mammals and humans are stacked on top of evolutionarily older parts which originally developed in reptiles. These rather different parts of the human brain need to function as a well-integrated whole. Such integration is achieved as the brain not only registers external stimuli, but also generates a lot of internal activity and responds to (and influences in an endless feedback loop) physiological processes throughout the whole body. This ongoing overall integration of the brain-body is vital if we are to navigate our natural, social, and technological environment in a productive and rewarding way.

Broadly speaking, the human brain can be seen as comprising three distinct parts—a paradigm suggested in the 1950s by neuroscientist Paul MacNeal who wrote of the "triune brain."[15] The brain stem sits on top of the spine and processes signals from different parts and organs of the body entering through the spinal cord. It contains centers which control basic physiological processes like breathing and the circulation of blood. This "reptilian brain" is also involved in the initial processing of touch, smell, and sound, and regulates key basic drives related to sex, social dominance, and territorial "ownership."

On top of the "reptilian brain" is perched the more recently evolved and complex "mammalian brain." It controls other basic biological processes like the maintenance of a steady body temperature and the release of stress hormones. It also generates various emotional reactions and regulates more complex social behaviors. The key centers involved in these processes, plus some older parts of the cortex, are often referred to as the "limbic system."[16] They provide instant emotional response to our complex physical and social environment. They also allow the development of the longer-term emotional attachments needed to motivate parents to take care of their offspring who, unlike newly-hatched reptiles, cannot survive on their own. Taken together, such emotional sensitivity and attachment provide the basis not just for individual survival, but also for the formation of stable family structures, as well as broader communal bonds.

These more primitive "brains" are wrapped in the neocortex. This thin but densely wrinkled neural sheet is the latest addition to the architecture of the brain. In humans, it contains some areas which are not so different from similar neural structures in other mammals, like the visual and motor cortices. But in the human brain the neocortex is much more developed. It underlies our capacity for language, thinking, emotional modulation, and a much more complex regulation of social behavior and even of basic drives.

Admittedly, this is an overly simplistic overview of the overall layout of the human brain. It does serve, though, to highlight the need for reliable integration of parts of the brain which are very different structurally and functionally. This integration is ideally achieved as lower and higher parts of the human brain are constantly involved in multiple processes of mutual activation and inhibition. As a result, the functioning of even the apparently more primitive "reptilian" and "mammalian" brains is modulated by signals coming from the neocortex and is thus quite different from ostensibly similar processes in animals. The activation of different parts of the neocortex, in its turn, is profoundly affected by signals coming from lower parts of the brain which convey information about external stimuli and physiological processes throughout the body.[17] So thinking and decision-making become infused with bottom-up affective responses and visceral sensations—which are, in their turn, modulated through top-down processing.

The orchestration of all this top-down and bottom-up signaling is thought to be the prerogative of part of the neocortex—the prefrontal cortex (or the "executive brain") which is also involved in abstract reasoning, planning, and the modulation of various emotional responses.[18] In its turn, however, the prefrontal cortex is activated by signals coming from the thalamus, a structure deep in the midbrain, which helps integrate various sensory inputs. According to neuroscientist Rodolfo Llinás, the role of the thalamus is so fundamental that it, rather than the prefrontal cortex, can be seen as the "conductor" which prompts different parts of the neocortex to come online in response to different stimuli.[19] The prefrontal cortex is also activated by dopamine synthesized in "reptilian" midbrain structures in response to (or anticipation of) various kinds of stimulation and to visceral signals coming through the brainstem. This intricate interplay between the neocortex and sub-cortical structures needs to be finely synchronized if the human brain is to function properly. The existence of these multiple feedback loops in the brain makes the task of understanding its functioning exceedingly complex.

The intricate hierarchical integration of the brain is made possible by the development of dedicated pathways which serve to tie the different parts together. Neuropsychiatrist Peter Whybrow refers to these as "information superhighways" since they carry multiple signals between spatially and functionally removed parts of the brain. These pathways contain relatively few neurons whose "cell bodies are rooted in the brain's stem and their long axons spread upward like the branches of a tree to effectively connect the emergency systems of the reptilian brain with the limbic system and the new cortex."[20] Signals traveling along these neural "superhighways" trigger the release of many neurotransmitters in different parts of the brain. The most important among these chemical messengers are perhaps glutamate, serotonin, norepinephrine, and dopamine. Changes in their concentrations profoundly influence patterns of activation in affected brain areas.

Among these chemicals, the roles of norepinephrine and serotonin seem less complicated—if anything related to the workings of the brain can be uncomplicated. While norepinephrine activates the neurons which absorb it, serotonin helps calm down such neural excitation. Dopamine, however, has more ambiguous chemical effects (often combined with those of glutamate), and its synthesis is affected by other neurotransmitters. It has long been associated with the experience of pleasure, or, more broadly, the processing of various "rewards." This perceived link was once demonstrated by multiple experiments which measured the release of various brain chemicals in response to different kinds of sensory gratification in lab animals—from the consumption of tasty food or drugs to mating. Those early experiments indicated that a dopamine squirt in the brain is normally experienced as intensely pleasurable. They therefore prompted neuroscientists to identify an associated "pleasure" or "reward center" in the midbrain—a part of the brain stem containing nodes of dopaminergic neurons.

For several decades, this model of the release and effects of dopamine seemed quite straightforward. That conventional view, however, was challenged by some neuroscientists.[21] One of them, Jaak Panksepp, started to wonder if lab rats had a real "reward" center in their brains, as the prevailing theory of the day stipulated. In the experiments he conducted back in the 1970s, animals with elevated levels of dopamine in their brains did not seem to experience much "pleasure" or "reward" as conventionally understood. Instead, they appeared restless and agitated. Panksepp also noted that the highest concentrations of dopamine in rat brains were reached when the animals anticipated, or were surprised by, rewards. Rewards which rats had learned to anticipate, on the other hand, triggered much weaker dopamine surges.[22] Other neuroscientists made similar observations putting into question the narrow association of dopamine release with pleasure. In the 1980s, Kent Berridge conducted experiments which led him to conclude that "wanting" could be distinguished (and potentially separated) from "liking"—the former mediated by dopamine, and the latter by opiates.[23]

To make sense of such observations, Panksepp felt he needed to reconceptualize the role of the dopamine system in the brain. He decided that instead of referring to "reward," it would be much more appropriate to speak of a "seeking system." This designation would indicate that the associated neural centers and pathways of that system seemed to provide the motivation animals needed if they were to engage in active exploration of their environment—which is essential for finding food and other resources, as well as potential mates. Dopamine, however, is also released in response to startling noises, other strong stimuli, psychological stress, physical pain, etc. It thus seems to play a broader role in generating motivation to pursue what enhances chances for survival and procreation, and avoid what diminishes these. For this reason, what Panksepp has dubbed the "seeking system" could be more broadly designated as the brain's "dopaminergic-motivation system."

As it turns out, this motivation system also underlies the kind of learning which can ensure future access to similar life essentials and—in humans—much more complex problem-solving on the basis of past experiences.[24] It can, then, be seen as generating the impetus for novelty seeking, curiosity, sensitivity to mere cues, and the overall thirst for the knowledge animals and humans need in order to navigate in a successful and "rewarding" way their natural and social environment. In Jonah Lehrer's words, such desire for knowledge "begins as a dopaminergic craving, rooted in the same primal pathway that also responds to sex, drugs and rock and roll."[25]

This realization leads to an obvious question: what can trigger the kind of dopaminergic craving which won't be satisfied by more direct sensory gratification; but would instead provide motivation for the pursuit of knowledge, particularly the kind of knowledge that is removed from immediate personal experiences? Resolving this conundrum requires a more detailed look at the way the dopaminergic motivation system facilitates the integration of higher and lower parts of the brain.

When different sensory stimuli are processed in the brain, the resulting neural signaling takes two routes. One activates immediately the amygdala, a pair of almond-shaped neural nodes which are a vital part of the limbic system. When sufficiently strong, such direct activation of the amygdala can set off immediate behavioral responses—as when we pull back from a hot plate or jump away from the projected route of a car racing toward us. Signals from the amygdala also set off the overall fight-or-flight response and the release of stress hormones and other chemicals throughout the body, a process which also affects the brain.

The other route of sensory processing passes through the prefrontal cortex. There, incoming signals are automatically cross-referenced with relevant associations, impressions, and—in humans—ideas called forth from memory. On the basis of this synthesis, the prefrontal cortex makes a more complex assessment of the stimuli. It sends signals to other parts of the cortex which influence the formation of overall sensations and perceptions. Signals from the prefrontal cortex also reach the amygdala, prompting it to quieten down in the absence of real danger, or once the danger has passed.

The intensity of the amygdala's initial activation provides an instant, automatic assessment of imminent threats. The amygdala, working in synchrony with the insula (a brain center recruited in the processing of basic and social emotions)[26] and parts of the prefrontal cortex, is also involved in a broader emotional "tagging" of sensory signals or events. These are thus marked with positive or negative "valence" with different intensity.[27] Such affective tagging can be influenced by signals coming from the prefrontal cortex which reflect broader experiences and expectations. Examples would include the stronger pleasure we feel when we sip more expensive wine; or the richer (or more conceptual) perception of a painting by an art critic.

Signals coming from both the amygdala and the prefrontal cortex activate dopaminergic neurons in the midbrain, a key part of the motivation system. In response, these neurons start releasing dopamine. Neurons in the midbrain "project" long axons (which form some of the "superhighways" Whybrow describes) to the prefrontal cortex, the amygdala, and other brain centers. Dopamine thus carries a chemically encoded motivational message which, in high concentrations, can produce a euphoric or even manic sensation. This neural "high" indicates that some bits and pieces of information may have greater importance for the organism, and are worthy of attention and storage in long-term memory.[28]

This is, very crudely described, the physiological basis of the drive to explore the environment and seek out vital resources, mates, and other "rewards," which can include the joy of mutual grooming and play. Inhibitory signals from the prefrontal cortex can then help calm down the amygdala and dopaminergic neurons in the midbrain, giving the brain and the whole organism a much needed respite from stimulation and emotional excitation. Such inhibition of affective arousal which is no longer appropriate or necessary is an essential function of the "executive brain."

I wish I could provide a more evocative description of these intricate processes. A more light-hearted or metaphorical account, however, would not be true to the spirit of contemporary neuroscience—so bear with me for a few additional technical details. The interplay of the dopaminergic motivation system, the amygdala, the "executive brain," and other neural networks is also affected by the release of many different substances. In addition to changing levels of neurotransmitters and hormones, tiny molecules of various chemicals called peptides are also released. These are secreted under the influence of affective arousal. They dock at multiple receptors on cells throughout the body, modulating their overall functioning. Emotions thus become whole-organism responses interlinked with brain activation in a never-ending feedback loop.[29] Another loop is created as the electromagnetic fields generated by the whole brain, and maybe the heart,[30] influence patterns of neural firing and cell activation throughout the brain and body.

Signals coming through neural pathways from internal organs play a similarly critical role in the overall orchestration of neural, mental, and behavioral responses. Among these, signals sent from and to neurons located in the heart[31] and gut,[32] and processed through the insula, are particularly important. Also, high levels of physical activity seem to stimulate the formation of new neurons, and to provide higher levels of energy within existing ones.[33] Such ongoing mutual activation, inhibition, and feedback contributes to the formation of instant and longer-term, emotionally colored assessments of the significance of new experiences, events, facts, and ideas. These complex affective-visceral responses are key to learning, as they help tag some impressions as worthy of storage in long-term memory.[34]

Such storage depends on the formation and reinforcement of synaptic connections. While dopamine seems to provide the motivation for repeatedly seeking the experiences triggering such neurophysiological changes, the release of norepinephrine and glutamate in the brain appears to directly facilitate the formation of new synaptic connections. This process is also assisted by other substances released in the brain as a result of emotional arousal. Meanwhile, epigenetic changes influence the synthesis of the proteins involved in the formation of long-term memories.[35] Norepinephrine, the brain twin of the stress hormone adrenaline, also increases the overall excitability of neurons, including dopaminergic neurons in the midbrain. As already noted, all these interactions and modifications, in their turn, are not merely a series of complex responses to external stimuli. They represent mental models influenced by flashes of relevant recollections and associations related to past experiences.[36]

This complicated and delicate neural dance, involving multiple brain centers, pathways, networks, chemicals, and electromagnetic waves and fields (and overall brain-body integration) is at the heart of our biological and social being. In a well-tempered brain latched onto a healthy body, such ongoing neural processing provides us with information regarding potential threats and opportunities, and with appropriate motivation to explore our environment, acquire new knowledge, and form lasting attachments. There is, however, a potential danger. The dopaminergic motivation system can be easily hijacked by powerful stimuli whose processing may not be entirely conducive to our personal thriving, and even to our longer-term survival.

8

A Fatal Attraction

A few years ago, AUBG tried to outsource its loss-making dining services. The company selected to take them over thought the best way to win the hearts and minds of the student body, faculty, and administration was to place several large TV sets along the walls of the canteen. Several months later the university needed to take back control of the struggling operation. The TV sets, however, remained as a legacy of the short-lived experiments in Zeitgeist chasing. They haunted the AUBG population, or at least me, for a few years until dining services were moved to a shining new building. Each time I went to the canteen to have a meal, I found it extremely difficult to keep my eyes off the nearest screen. Even if I was talking to someone, my eyes kept darting back to the moving images.

As I glanced at the screens, I was reminded of a scene from the movie *Limitless*. In it, the main protagonist, a struggling writer, starts taking an illegal drug which tremendously increases his brain power. Naturally, he decides to use his newly acquired supernatural ability to strike it rich on the stock market.[1] His performance is predictably impressive, so he is hired by Warren Buffett's evil twin to assist in the execution of a giant corporate merger. At some point, though, the effects of the drug begin to wear off, and the newly hatched capitalist shark's mind and eyes start to drift. His boss immediately notices his waning concentration and snaps derisively: "Oh, don't tell me you are one of those guys who are easily lost when there is a TV screen in their vision field."

Unlike me, the writer-turned-analyst had a good excuse—his attention was caught by coverage of a grisly murder he could have committed without being aware of it. But the truth is, you do need superhuman powers of concentration and inhibition if you are to resist the pull of a bright TV screen. There is a fairly straightforward physiological explanation for this phenomenon—which once made television into the paradigmatic centerpiece of modern living.

Our attention is easily attracted by any novel stimulus or moving object. These activate instantly an automatic "orienting response" in the brain. The eyes latch onto anything that intrudes into, or moves within, our visual field since any such object may be a potential source of food or mortal danger, or a possible mate, friend, or enemy. This reflex prepares the organism to deal with potential opportunities or hazards.[2] It, or anticipation of related stimuli, activates the amygdala.

The intensity of this activation provides an instant, automatic emotional "tagging" of sensory signals or events which may have high survival value for the organism. As the example with the almost irresistible TV screen suggests, however, such a response can sometimes be misleading. The activation of the amygdala, and the related release of dopamine in the brain, can be triggered by empty but powerful stimuli—for example, loud noises and bright flares that are merely startling.[3]

Television is a prime example of this tendency. We can be easily mesmerized by its bright, flickering images for several hours every day—an "activity" which does not exactly enhance our chances for survival or procreation.[4] How can we, then, resist the pull of such "chewing gum for the eyes"? To achieve this feat, we must be able to inhibit the automatic "orienting response" triggered by ever larger and brighter TV screens, and redirect our attention to more important stimuli and pursuits.[5]

This activation, monitoring, and redirection of attention depends on a delicate balance between emotional arousal and inhibition in the brain and the whole organism. Some neuroscientists place at the core of this balance the interaction of two neural networks I have already touched upon. One of them, which they refer to as the brain's "impulsive system," is built around the amygdala. The other one, called the "reflective system," is centered on the prefrontal cortex and thus largely overlaps with the "executive brain" described earlier.[6] The orienting response activates the impulsive system, which immediately directs one's attention to potential threats or "rewards." The reflective system, on the other hand, monitors this activation and generates a more complex assessment of the importance of the stimuli or events which have turned it on. This assessment eventually leads us to either persist in paying attention or switch attention to a more worthwhile object.

This is roughly how things should work ideally, but they often do not. The balance between the impulsive and reflective systems in the brain can easily be disrupted. The activation of the impulsive system functions according to a rough winner-take-all principle. This means that stronger signals tend to drown out weaker ones. For example, a loud MTV-style video clip can easily distract us from observing the blossoms of a plant next to the TV screen; or the attraction of a piece of chocolate cake can make a slice of cucumber seem unappealing. Also, the anticipation of stronger sensations can cancel the attraction of more feeble ones.

The more powerful signals, or the keen anticipation of associated stimuli, can thus become the triggers for the release of dopamine which heightens arousal in the brain and body. Under normal circumstances, the reflective system in the brain serves to place such powerful excitation into perspective and keep it within healthy limits. But if the underlying dopamine-induced "high" is particularly strong, signals generated by the prefrontal cortex may be too weak to achieve this vital task. As a result, the dopamine system can be hijacked by large doses of intense but pointless stimulation.[7]

The dangers of such potential neural dysregulation were first demonstrated by experiments with lab rats conducted almost half a century ago. The hapless animals were taught to press a lever in order to self-administer stimulant drugs or deliver electrical signals to electrodes implanted in the dopaminergic centers in their brains. In addition to the rush induced by the resulting dopamine squirts in their brains, such stimulation also induces the secretion of endorphins, the brain's own opiates which can make the dopamine-induced excitation feel pleasurable. Once the animals entered that euphoric state, they typically kept pressing the lever to the point of full exhaustion and even death. As they engaged in an apparent orgy of self-stimulation, they disregarded easily available food, drink, and even mating partners.

Of course, humans who find themselves in similar situations should know better. The prefrontal cortex in the human brain is much larger and functions at a whole different level. It is thus capable of sending stronger signals intended to curb potentially unhealthy impulses. But even in humans the top-down signals coming from the prefrontal cortex can be overwhelmed by the powerful rush and excitement triggered by higher concentrations of dopamine.

Humans do not normally have electrodes implanted in their brains which can deliver appropriately targeted impulses to their dopaminergic circuits. But they can resort to other forms of self-stimulation, for example by taking powerful drugs. As in lab animals, the strong dopamine response provoked by these can incapacitate the brain's reflective system. As this happens, the impulsive system takes over, and the reflective system is effectively shut down. As a result, receiving another dose of the drug turns into an obsession. This fixation is reinforced by other chemicals released in the body.

Drugs can thus become an object of intense desire and craving. The end result is an obsession that outweighs the motivation to pursue any other goals.[8]As a consequence, drug addicts typically experience an erosion of willpower. They come to be driven by strong "nonplanning" and "cognitive impulsivity." They tend to pursue immediate sensations, and to jump to conclusions in the absence of clear clues. Any object or place associated with drug use can trigger an intense craving, sometimes years after overcoming the related addiction.[9]

Over the longer term, such addiction to stimulant drugs rests on a particularly intractable vicious circle. The brain is gradually habituated to the chemical effects of drugs. The secretion of dopamine in the dopamine system is decreased and neurons in other brain areas lose some of their dopamine and opiate receptors. As a consequence, they become less sensitive to the effects of that substance. The intense sensations a drug induces then tend to fade. The only way to restore sensations to their previous intensity is to take larger doses of the drug. Chronically elevated dopamine levels resulting from such substance abuse can cause neurons in the prefrontal cortex to wither and even die as a result of oxidative stress—a process which reduces the power of the "executive brain" to function properly.[10]

Such neural modifications contribute to a self-defeating spiral familiar to heavy drug users. Beyond a certain point, even much higher doses can no longer provide quite the same sensation. As this happens, the responsiveness of addicts to non-drug-related stimuli is dampened as well. Their daily existence is robbed of almost any other pleasure or excitement which could be provided, for example, by a walk in the park or a pleasant conversation. As a result, the lives of addicts feel a lot duller. At the root of this unfortunate tendency is a physiological malfunction—the endorphins released in the brain and body in response to potentially pleasurable experiences find fewer available receptors to dock to. This progressive numbing of the senses, erosion of willpower, loss of a larger perspective on life, and the resulting desperation can lead addicts to a downward spiral of self-defeating sensation-seeking and sometimes untimely death.[11]

As the similar effects of electrical self-stimulation and

stimulant drugs illustrate, any kind of strong neural activation which triggers a dopamine rush and related cravings can disrupt the delicate balance between the impulsive and reflective systems in the brain. Such progressive neural dysregulation predictably leads to heightened impulsivity and to decreased motivation to delay gratification and resist all sorts of unhealthy temptations. Brain-imaging experiments have demonstrated that the balance between the impulsive and reflective systems in the brain can be similarly disrupted by highly palatable food, sex, pornographic images, money, and other powerful stimuli.

The processing of money in the brain is particularly indicative, since money offers a much more abstract "reward" than any chemical substance or directly pleasurable sensation. As some curious experiments have demonstrated, not just viewing or counting, but merely thinking of money can trigger strong affective signals in the amygdala and the release of dopamine in the midbrain. As with drugs, the brain can become habituated to the ensuing rush associated with the anticipation of high monetary rewards. Individuals can thus become stuck in pursuing ever larger and more elusive monetary payouts. In the process, they tend to become less sensitive to many other pleasures in life.[12]

These effects are particularly acute in compulsive gamblers. As the highly perceptive self-described gambling addict I mentioned earlier confesses, big wins and losses at casinos would initially trigger in him powerful excitement or desperation.[13] These reactions were all too natural, since unpredictable rewards produce the most intense dopamine rush. But gradually the sensations grew fainter. As sums on the order of 14,000 dollars drifted in or out of his hands during gambling all-nighters, he no longer felt thrilled or shocked. After years of gambling, he came to a point when he felt nothing. Worse, he did not have a clear sense as to what he should be feeling under such circumstances. Despite that stupor, he kept traveling to distant casinos and spending endless nights gambling, while his ostensible girlfriend was waiting patiently for him in their hotel room.

This apparent affinity between compulsive, often reckless gambling and recognized addictions has long been noted. For this reason, "Gambling Disorder" is the only recognized "behavioral addiction" included in the new edition of the American Psychiatric Association's Diagnostic and Statistical Manual (DSM-5).[14] "Internet Gaming Disorder" is also listed there, though only as a "condition for further study." But once such concessions are made, other problematic behaviors must sooner or later follow since there is no conceptual reason for keeping them out.[15] As already noted, internet addiction (including online gaming) has already been recognized as a mental disorder in China, South Korea, and other countries.[16]

In fact, the similar effects drugs, food, sex, pornography, gaming, gambling, and other compulsive behaviors can have on the brain (and, by extension, on the body) raise an interesting proposition. It seems that the brain uses a common neural "currency" to assess various potential rewards, threats, and gratifications.[17] As the abstract nature of astronomical monetary rewards demonstrate,[18] the objects or stimuli whose value is estimated in this neural currency do not need to be tangible. They can be akin to the thrill or excitement provided by any strong sensation—like the ones generated by an "epic win" in a video game, daring stunts in extreme sports, shopping sprees, the acquisition of new piercings or tattoos, and other similarly exhilarating experiences.

Among these activities, there is one which has acquired an increasingly central place—seeking out and processing novel information. As already noted, dedicated circuits in the brain can be activated by strong sensory stimuli which do not point to potential rewards beyond the dopamine rush they trigger. As a result, the motivation system can be hijacked even by stimuli which carry little meaning or significance. This danger has been well captured by journalist Emily Yoffe who refers to Panksepp's contribution to the study of the effects of dopamine on the brain. She observes a disheartening resemblance between human and rodent self-stimulation. In her words, "while we tap, tap away at our search engines, it appears we are stimulating the same system in our brains" which those lab rats sought to tickle with abandon.[19]

In recognition of this affinity, the excessive, often compulsive seeking and accessing of information from the internet has been dubbed "infomania," a form of dependence which can be "worse than marijuana."[20] The tendency to stay online for hours every day surfing, communicating, and cultivating an attractive "e-personality" has in itself been identified as an addiction.[21] Increasingly, such virtual connectedness is becoming the water in which we all swim. It has even crowded out many older forms of self-indulgence in our, and our children's, lives.

Playing ostensibly pro-social games like *The Sims* or *FarmVille* (or "stupid" ones like *Angry Birds*[22]), posting innocent comments and photos on Facebook or Instagram, browsing non-pornographic material online, or switching between multiple apps may seem far removed from the highly addictive experiences described above. Yet they are driven by similar impulses and have similar effects on the brain, and particularly on the prefrontal cortex.[23] According to Nora Volkow, the flamboyant director of the National Institute for Drug Abuse in the United States, there is no substantial difference between substance abuse and the various "behavioral addictions" increasingly coming under scrutiny.[24] At rock bottom, they are all driven by a chemical craving at the cellular level, though compulsive behaviors may not exactly meet the strict criteria required for designating them as clinical disorders.[25]

Still, many lay observers and commentators remain unconvinced that labeling a widespread behavior like "excessive" internet or gadget use as a disorder can be helpful. Virginia

Heffernan again provides a symptomatic example. In her capacity as a "social media" critic, she was not particularly bothered by the confessions of a self-described internet addict.[26] The young woman acknowledged she had a tendency to lose herself clicking from one hyperlink to the next, sometimes going to bed in the early hours of the morning. Yet, she appeared to Heffernan as merely "a bright, self-effacing, religious young woman who keeps student hours and prefers logic games, jokes, graphic novels, trivia quizzes, music, Victoriana and socializing on Facebook to prefab pop bands."[27] Heffernan would probably be unfazed by the apparently obsessive behaviors depicted in documentaries like Douglas Rushkoff's *Generation Text*, Anderson Cooper's *#Being13*, or Delaney Ruston's *Screenagers*.

Heffernan's predictably blasé attitude is supported by some addiction researchers and neuroscientists. They point out that seemingly problematic behaviors like sex or information seeking are natural, and do not trigger the release of nearly as much dopamine as the consumption of powerful drugs.[28] But the effects of even apparently innocuous stimuli on the brain can be much more sustained as compared to some drugs. Such potentially problematic behaviors may come to occupy most of the waking hours of affected individuals. This cannot but have lasting effects on brain wiring and patterns of overall emotional arousal.

In the case of information seeking and consumption, those effects were first demonstrated in a 2007 brain imaging study by neuroscientist Gary Small.[29] He wanted to compare the patterns of neural activation in experienced internet users and in complete novices who—incredibly—had never surfed the web. Quite surprisingly, Small found that all it took to induce observable changes in the novices' patterns of brain activation was several hours of internet browsing. Their brains had started to fire in ways similar to those of experienced users.

Other researchers have similarly confirmed that excessive

online gaming induces changes in cortical areas and in deeper brain structures. These modifications may be adaptive in the world of online gaming and other areas requiring rapid response and precise hand-eye coordination.[30] But they can also create problems related to impulse control and sustained attention to weaker stimuli. Such problems are likely to be even more acute in the case of ostensibly pro-social internet games, which have gained popularity courtesy of ubiquitous internet-connected devices, and the "social media" centered on user-generated content and regular peer-to-peer "sharing" and communication. Such compulsively pursued activities seem to offer an addictive combination of adventure, socializing, info-grazing, and other "micro-rewards" which can induce neural modifications similar to those observed in the brains of drug addicts.[31]

As in the case of monetary rewards, an "infomanic" lifestyle can have an additional and more insidious effect—a partial desensitization to non-electronic stimuli. If we spend countless hours seeking instant feedback from screens of various sizes, our nervous system will be partly habituated to the dopamine rush induced by such sensory and emotional overstimulation. As a result, we will need an ever more intense and prolonged exposure to the novel or constantly refreshed images and sensations generated by virtual immersion. We will seek to experience the previous excitement, but will get no lasting satisfaction.

Fortunately, few among us will succumb completely to the influence of such virtual thrill seeking or die of an overdose like those hapless lab rats or gamers in Eastern Asia. But as a result of chronic exposure to electronically generated or mediated feedback, the non-virtual, allegedly "real" world may, indeed, lose much of its urgency and relevance.[32] Non-electronic sources of information may begin to pale and seem unexciting compared to virtual reality and interaction. In fact, cultural critic Thomas de Zengotita thought this shift in perception and comprehension had already become the new normal (in the United States at least) at the time of the 9/11 attacks.[33]

Such a shift will most obviously affect the fate of the written page. As Carr and other internet skeptics have noted, habitual online info-grazing may lead us to a point where potentially interesting and important texts fail to provoke the intense and sustained engagement associated with "deep reading."[35] By making us less sensitive to all sorts of stimuli and experiences, submersion in the "apptwinstagoocebook" universe may even turn us a bit more obtuse and indifferent in general—making us blasé about these very effects of our "infomanic" existence.

9

Dumb and Dumber

Armed with all these new insights, I could now get a theoretical handle on some of the worrying cognitive trends I was observing in my classes. Why were the two articles or book chapters I was usually assigning in upper-level classes—which, until a decade ago had seemed more or less manageable—beginning to look like an overwhelming burden? Why were so many students losing interest in the social and political world I and the authors I assigned found so compelling? Why were the majority of students having difficulty accumulating the background knowledge and developing the conceptual frameworks needed to grasp and store in long-term memory new issues and ideas? Why were they finding it even harder to "transfer" concepts and knowledge acquired previously to new courses and contexts? Why was the development of cumulative knowledge and holistic understanding of larger issues—which is the *raison d'être* of liberal arts education—beginning to look less realistic? Why were so many students failing to develop an intuitive sense of English grammar, syntax, word formation, and pronunciation after four years of reading and writing in English, and—in many cases—time spent working or studying in the United States? I thought I had an elegant theory explaining it all.

A new study published in early 2011 under the ominous title, *Academically Adrift,* seemed to corroborate all these observations.[1] It decried in strong terms the "limited learning" allegedly taking place in American higher education. Its authors, sociologists and education experts Richard Arum and Josipa Roksa, based that conclusion on an analysis of results from the Collegiate Learning Assessment test taken by 2,300 students at a cross-section of American colleges and universities.

The centerpiece of that test is a "performance task component" in which students have 90 minutes to draft a solution to a practical task (for example, what would be the best way to reduce pollution in a particular area?) on the basis of several documents providing relevant information. The idea behind this task is to assess some higher-order cognitive abilities, like critical thinking, analytical and integrative reasoning, and written argumentation. As Arum and Roksa crunched the test results, they reached some grim conclusions. They found that 45 percent of the students they tested did not demonstrate significant improvement in learning during the first two years of college; and 36 percent showed no such progress over four years of college education. Moreover, the improvement the majority of students did show tended to be very modest.

Arum and Roksa singled out as the main explanatory variable for these disturbing results the insufficient rigor of most college courses and the limited amount of time students spent studying. They cited results from surveys indicating that 32 percent of students each semester did not take a single course requiring more than 40 pages of reading per week, and roughly half did not take a course in which they needed to complete more than 20 pages of writing for the whole semester. As they took such undemanding courses, students spent on average only 12-14 hours a week studying. Much of that time they studied in groups, which Arum and Roksa did not find very productive. Meanwhile, students invested much time in extracurricular activities which did not seem to produce measurable educational returns. The authors found statistically significant correlations between the rigor of the courses students took and the time they spent studying alone, and improvements in learning.[2]

When *Academically Adrift* came out, it (and the swirl of articles and press releases generated by the accompanying publicity campaign) created a firestorm. More upbeat commentators dismissed the study as "statistically adrift" or questioned its reliance on a supplemental test which few students had the incentives to take very seriously.[3] But there was also a broad sense that even if the numbers weren't exactly right, the study had indeed pointed to declining academic standards. Meanwhile, other observers had already offered an explanation similar to the one Arum and Roksa put forward. For example, the 2005 PBS documentary *Declining by Degree* included interviews with apathetic students and frustrated professors and college administrators.

Leave Them Kids Alone!

As a typical example of reduced course requirements, *Declining by Degree* introduced an economics professor at a large state university who had made the textbook for an introductory course optional. Instead, he required students to read breezy articles from *The Economist*. He acknowledged that he would be given or denied tenure mostly on the basis of his published research, not his teaching.[4] The documentary also profiled students, however, who lacked any curiosity or desire for learning, and were not particularly embarrassed to admit this on camera. College administrators interviewed for the documentary alleged that students and faculty at research universities had entered into a mutual non-aggression pact, agreeing not to be overly demanding of each other.

The problem that the majority of both professors and students would rather be "elsewhere" is highlighted in a sermonizing article by Mark Edmundson.[5] His poignant message is addressed to the class of students entering college in 2011, or maybe to the very few among them who can realistically be expected to read anything like the text he wrote. Such observations regarding the priorities of most faculty members, though, may be less relevant to many lower-tier universities and colleges which place a stronger emphasis on teaching. They definitely do not apply to my place of work, AUBG, where expectations for research are modest, with teaching clearly instated as the main institutional priority.

Such explanations seem to corroborate and offer a plausible explanation for many surveys which have found that American students, indeed, have shaky knowledge about almost any social or political topic. One of the largest samples of such data is collected in English professor Mark Bauerlein's book, *The Dumbest Generation*.[6] Here are just some of the most striking findings he quotes. In a 2004 study, only 28 percent of 18-to-26-year-olds could identify William H. Rehnquist as the chief justice of the United States; 26 percent "could name Condoleezza Rice as Secretary of State, and … 15 percent knew that Vladimir Putin was the President of Russia."[7] Meanwhile, students who lacked this kind of knowledge, or even had trouble listing the three branches of government, recited with ease (and, presumably, stronger enthusiasm) the names of the Three Stooges and of the latest American Idol.

Calling the "Millennials" *in toto* "the dumbest generation" may be a bit harsh and insensitive. This offensive label is calculated to evoke an unflattering comparison to what news anchor Tom Brokaw once dubbed "the greatest generation." Still, the massive amount of data Bauerlein has collected does seem to demonstrate that even the Millennials who have gone on to pursue a college degree have learned much less than expected about anything unrelated to the popular youth culture of the day (or their own narrower sub-culture).

Bauerlein has attributed the dearth of political knowledge

among college students and recent graduates largely to the influence of information technology. In his view, electronic communication allows the majority of the Millennials, the proverbial "digital natives," to encase themselves in a generational cocoon. Within it, they remain cut off from adult frames of reference and concerns. Engrossed in incessant peer-to-peer contact and a world of immediate realities, they can hardly be bothered to grasp and retain knowledge about larger social issues. These all seem to them distant, irrelevant, and instantly boring.

Bauerlein's theory does seem intuitively plausible. Yet, the trend he describes seems to predate the spread of the internet and constant access to "social media." For example, in a study conducted over two decades ago college students gave 10 million as a mean estimate of the number of Muslims in the world. Close to 70 percent of them thought there were more Jews than Muslims in the world. Incidentally, 40 percent of participants in this study were enrolled at Berkeley.[8] According to another study, published by a conservative think tank a decade ago, students at some prestigious universities seemed to have lost part of the knowledge they had acquired in civics and history classes in high school.[9]

Bauerlein's focus on the content of the peer-to-peer communication facilitated by the internet may, in fact, be slightly misleading. Marshall McLuhan taught us as much over five decades ago. Proclaiming that "the medium is the message," he argued that it is the nature of the channel of communication itself that largely shapes the reception of any message.[10] And this happens at a broader societal level, not just at the level of individual perception and understanding. McLuhan defined the media broadly to include "any technology whatever that creates extensions of the human body and senses, from clothing to the computer."[11] Long before the rise of the internet, he concluded that "electromagnetic technology" was giving rise to a completely mediated social environment. It was thus causing an information overload which numbed the senses, inducing exhaustion of the central nervous system and general bewilderment.

McLuhan was particularly fascinated with the explosive spread of television and the effects of the new medium on children and teenagers. In fact, he attributed the generation gap which suddenly opened up in the 1960s largely to the effects of the new but already pervasive medium. To illustrate this often overlooked point, he commissioned an iconic mural called "Pied Pipers All." The painting depicted stylized human figures writhing in a psychedelic dance against the background of a bright TV screen. McLuhan wanted that image to represent vividly the way in which television was mesmerizing the young and leading them away from adults—as the proverbial "pied piper" had once done with the children of a medieval German town. He attributed this cultural upheaval and the related "generation gap" to the influence of television on the human brain.[12]

McLuhan is often dismissed as a flamboyant, media-savvy media guru who had perfected the art of the sound bite—but whose cryptic pronouncements held little analytical value. As we shall see later, some of his conclusions regarding the impact of television on society were, if anything, overly optimistic. But his overall idea that massive daily exposure to a new screen-based communication technology must have a profound impact on the nervous system, perception, and social functioning seem truly prophetic. In the 1970s, writer and birdwatcher Marie Winn raised even stronger concerns, calling television "the plug-in drug."[13] These early warnings later received support from some findings in neuroscience. On the basis of these, psychologist Mihaly Csikszentmihalyi and media researcher Robert Kubay eventually concluded that television was, in fact, addictive—in the most literal sense of the word.[14] TV addiction has not yet been recognized as a clinical diagnosis, but the basic idea that it causes unhealthy changes in brain wiring and activation is probably sound. Needless to say, these changes have been tremendously intensified by the spread of the internet and video games, the rise of "social media," and the proliferation of hand-held "platforms" allowing constant access to these (and loaded with multiple apps and games).

The extent of this digital immersion, particularly among adolescents whose brains are more easily affected by all kinds of influences, now defies the imagination. A study completed in 2009, before smartphones had become ubiquitous, found that American children and teenagers aged 8 to 18 were spending on average 7.5 hours a day interacting with one or more electronic screens. That number did not take into account multitasking, an hour and a half spent texting, and a half hour of talking over a cell phone. The title of *The New York Times* article which reported those findings in 2010 said it all: "If Your Kids Are Awake, They Are Probably Online."[15] Two similar studies had previously been done in 1999 and 2004. The second one had found a significant increase, and its authors "had concluded ... that use could not possibly grow further." They were in for a surprise. According to the 2009 data, "young people's media consumption grew far more in the last five years than from 1999 to 2004, as sophisticated mobile technology like iPods and smartphones brought media access into teenagers' pockets and beds."[16]

Digital Divide 2.0

According to a study published in 2010, children whose parents lacked a college degree (which is taken as shorthand for lower socioeconomic status) were spending on average 90 minutes more per day exposed to screen-based media compared to families where at least one parent had a higher-education degree. Since 1999, that exposure had increased by four hours and 40 minutes and three hours and 30 minutes for the two groups respectively (the study double-counted hours spent multitasking). On the basis of those numbers, it seemed that efforts to bridge the proverbial "digital divide" had the paradoxical effect of creating a new gap with even more troubling implications. The *New York Times* article which reported these findings focused on the time children and teenagers were wasting online, but the direct impact of all that time spent staring at screens of various sizes may be a lot more profound.[17]

We can safely assume that over the past few years this trend has continued unabated, and may have even picked up speed with the explosive proliferation of smartphones and tablets. Another study conducted in the now distant 2010 found that 94 percent of American high school and college students had Facebook profiles. They spent on average 11.4 hours per week logged into the website. Seventy-eight percent of them accessed it using their mobile phones.[18] It has become increasingly common for schoolchildren and teenagers to spend even recess glued to hand-held screens, watching video clips or interacting through Facebook. Once at home, they commonly multitask: chat with friends, watch videos or check out newly posted pictures, scroll through Facebook, listen to music, etc., while they are doing their homework.

With all these data in mind, it can only be expected that the problems troubling the allegedly "dumbest generation" of "academically adrift" college students start earlier. According to a 2006 study, just over a quarter of American teenagers graduating from high school were "proficient" in civics. One third, on the other hand, did not have even "basic" knowledge in that area.[19] Results in U.S. history were even worse. Only 13 percent of students were proficient, and over half did not have even basic knowledge.[20] Though the reports emphasized some tiny improvements as compared to the previous studies done in 1998 and 1994 respectively (and the way "proficiency" is "operationalized" can be questioned), they still drew a fairly bleak picture.

It might be tempting to see such results as a uniquely American problem. When compared internationally, however, it seems that "on average, American youth perform fairly well in international comparisons of civic and political knowledge."[21] A study of the political knowledge and attitudes of young Europeans published in 2002 revealed a similarly disconcerting picture. Most of them "believed that citizens should vote and obey the law." They also expressed support for "social movement activities." But they showed little enthusiasm for "conventional political activities" like the "discussion of political issues."[22] Curiously, a recent study has found a striking degree of ignorance among Belgian teacher trainees. For example, "when asked which political ideology stood for the redistribution of wealth, higher taxes and state involvement, only one in two answered socialism." Also, "among final year teaching students involved in the study, one in three could not identify the United States on a map and almost half did not know where the Pacific Ocean was."[23] The problem of disengagement or intellectual and existential distancing from larger issues—despite recent protests and other forms of youth mobilization in many countries—thus seems to be part of a much broader syndrome.

The failure to develop a strong interest and to accumulate sufficient knowledge in such issue areas has gone hand in hand with another significant trend—lagging reading proficiency. A 2003 study already found an "inexplicable" decline in the reading proficiency of college graduates over the preceding decade. Only 31 percent could "read a complex book and extrapolate from it," as compared to 40 percent in 1992. Among graduate students, only 41 percent were rated as "proficient" in reading—a decline of 10 percent.[24] The same study also revealed another unexpected development—more college students demonstrated intermediate reading abilities. That result led the authors "to question whether most college instruction is offered at the intermediate level because students face reading challenges."[25] That suspicion was shared by Dolores Perin, a reading expert at Columbia University Teachers College. After sitting in many high school classes, she had become convinced that there was a seldom discussed but "tremendous literacy problem among high school graduates." As a result, "the colleges are left holding the bag, trying to teach students who have challenges."[26]

In recent years, 40 percent of incoming first-year students at American four-year colleges have needed to take at least one remedial class in reading, writing, or math.[27] It should then be hardly surprising that, as philosophy professor Carlin Romano has noted, "for too many of today's undergraduates, reading a whole book, from A to Z, feels like a marathon unfairly imposed on a jogger"—an attitude which has predictably brought about "the disappearance of 'whole' books as assigned reading" in college courses.[28] The proverbial "crisis in education" can thus be seen, at a more basic level, as reflecting primarily a crisis in reading.

Many studies suggest that over several decades reading for pleasure has declined considerably in the United States. As writer Caleb Crain has summarized the data, "we are reading less as we age, and we are reading less than people who were our age ten or twenty years ago."[29] He has also pointed to "indications that Americans are losing not just the will to read but even the ability." According to one study, between 1992 and 2003 the proportion of adults who qualified as proficient readers (who could, for example, compare the viewpoints expressed in two editorials) declined from 15 to 13 percent.[30] These are again not exclusively American problems. A study found that in the Netherlands reading for pleasure had similarly declined since the 1950s as TV viewing had increased. In the mid-1990s, college graduates born after 1969 were reading less than people without a college degree born before 1950.[31]

There must be many contributing factors to such a steady decline in reading, reading ability, and social knowledge. An increasing immersion in a virtual environment, however, is probably among the most important. The original "virtual reality" was created decades ago as television became truly pervasive and a ubiquitous babysitter. More recently, constant access to the web, video games, and numerous apps has provided the last straw (for now). Once serious reading turns into a boring obligation, the pull of virtual distractions is likely to become even stronger, creating a vicious circle that is all but impossible to break. This compulsion is likely to persist after teenagers enter college and beyond, contributing to pervasive learning problems.

10

...or a Neurosomatic Crisis?

These shifts in preferences and aptitudes have come as a bit of a surprise. Developmental psychologists tend to assume that brain maturation follows a "natural" path leading toward the acquisition of higher cognitive abilities—sophisticated reasoning, expert reading, broad knowledge, social competence, and purposeful self-discipline and planning. They see the human brain as exceedingly robust[1] or plastic in its capacity to adapt functionally to changing social and technological environments.[2] Its remarkable resilience can presumably be disrupted only by traumatic experiences like child abuse or rare genetic vulnerabilities. Suboptimal parenting or a generally unhealthy social and technological milieu, on the other hand, are rarely regarded as sufficient reasons for worry.[3] This line of thinking does not rest on much solid basis, other than the assumption that familiar skill sets and knowledge chests must be the result of a natural evolution. It may need to be radically rethought in light of the growing cognitive difficulties experienced by students at various levels.

With social modernization, humans have needed to adapt to a highly unnatural habitat: densely populated cities, overcrowded places of work and learning, tight daily schedules, an artificially induced sleep-wake cycle, an overall technological infrastructure which encourages sedentary lifestyles, a confusing array of lifestyle, temptations, and consumer choices, eroding social standards and hierarchies of authority, etc. Since the mid-20th century, children, adolescents and adults in richer countries have been exposed to a social and technological milieu which, with its growing complexity and stress levels, has become increasingly overwhelming. The sensory and social overstimulation provided by immersion in a screen-based virtual reality has added an additional layer to this overall picture.

The suspicion that modern societies and technological progress create a deeply unnatural and potentially unhealthy living environment is hardly new. It can be traced back at least to Rousseau and the Romantic poets and thinkers who followed in his footsteps. For a long time, their worries appeared premature and overblown. Up to a point, the human brain seemed capable of developing new abilities, adapting to the accelerating waves of technological and social change, and generating inventions and abstract models which helped accelerate further the whole process of intensifying and often disruptive social transformation. On the basis of this overall experience, it is tempting to still think that our brains and whole organisms are infinitely adaptable. This, alas, is a dubiously rosy assumption.

The biologically "modern" human brain developed over 100,000 years ago in the African savannah. Over a long prehistoric period, it was well adapted to the simple lifestyle of our hunter-gatherer ancestors. Succeeding stages of social evolution produced the first sparks of recognizable symbolic "culture;" ushered in agriculture, cities, and writing; and made possible the growth of ancient chiefdoms, empires, and city-states.

In the course of those momentous shifts, it turned out that the human brain did have much excess capacity and plasticity.[4] It could undergo massive reorganization, and many of its areas have been put to novel uses—a process which has helped humans, for example, learn to read and engage in abstract thinking and complex planning. Yet, there must be a limit beyond which the brain cannot successfully adapt, or adaptation would produce odd permutations analogous to the strange colors and protrusions of fish that have mutated to live in the depth of the ocean; to say nothing of the unhealthy neural adaptations associated with substance-dependence and behavioral compulsions. Alas, that limit may have already been reached.

Mastering our contemporary technological and social habitat and thriving in it strikes me as an almost impossible task. It requires a cognitive acumen and finely calibrated emotional control which can develop only under the most exceptional of circumstances.[5] These abilities can flourish only through utmost individual and institutional efforts, and are inherently precarious.

In this sense, we do not resemble the highly intelligent dolphins who can learn effortlessly to cooperate and communicate (apparently, calling each other by name)[6] in order to catch more fish or evade killer whales. Cognitively, we are more like the border collie who was able to learn the names of 1,002 different objects after years of training with her dedicated and infinitely patient owner, a retired college professor.[7] Socially, we are more like circus animals who have learned to walk on their hind legs and mastered complex tricks. We can also compare ourselves to those chimps that, after years of living with humans, have learned to use some sign language, drink from a bottle, and go to the bathroom or wear nappies. Such skills and knowledge require powerful motivation, persistent effort, and rigorous training. They are unlikely to develop, and would be superfluous, under any set of "natural" conditions. Such "domestication," even if necessary, may come at a price—as recognized by Sigmund Freud or Friedrich Nietzsche (who famously described man as "the sick animal"). It may also remain incomplete and can even be reversed[8]—a tendency Freud once thought was unfolding as World War I took its brutal course.

If an overall grasp of our social and cognitive predicament is insufficient to corroborate such premonitions, they can perhaps receive support from some experimental findings. Among these, the results from a series of rigorous studies conducted by German researchers over several decades are particularly striking. In the 1980s, they started to observe with surprise some dramatic changes in the neural responsiveness and behavior of children. German children were apparently becoming less sensitive to milder sensory stimuli, and would often register and pay attention only to "brutal thrills." [9]

The German researchers thought this decreased sensitivity was a result of the need of the brain to filter out much of the overstimulation it was receiving. They called it the "horn of plenty" effect. Children who exhibited this syndrome also easily tolerated dissonant information, without expressing the expected emotional discomfort and seeking to integrate the various bits and pieces into coherent patterns. A degree of sensory numbing thus seemed to go hand in hand with new deficiencies in conceptual processing and orientation.

The overall result of these changes in sensory, emotional, and cognitive processing was decreased appreciation for some fundamental features of the surrounding reality. That created a degree of detachment which Dutch trend philosopher Gert Gerken dubbed "the new indifference."[10] All those striking changes were apparently the result of incessant sensory overstimulation from a very early age. On the basis of those observations, the German researchers concluded that children born since the late 1960s had developed a "new brain" with distinct wiring patterns. The symptoms the German researchers observed, though weaker, looked similar to the dulled perception of "reality" induced by substance abuse or behavioral addictions. Ironically, that "new brain" is now the brain of the "digital immigrants" which seems to possess stronger integration and coherence than the brains of the upcoming "digital natives."[11]

German children are unlikely to represent some sort of weird exception in this area. Indeed, experts and teachers in the United States have observed a similar increase in impulsivity and generalized learning difficulties in children. In the words of Martha Herbert, she and her colleagues have started to see more children "with diffuse difficulties—not discrete learning disabilities where everything else is more or less intact, but difficulties spread across multiple cognitive, sensorimotor, social, and emotional domains."[12] Herbert made these observations back in 2000, citing statistics indicating that 17 percent of American children had "some kind of attentional or learning problem."[13] As I noted earlier, she thought such difficulties reflected changes in brain wiring induced by an overwhelming social and physical environment.

The difficulties of learners who experience similar symptoms and cognitive problems are, indeed, likely to be generalized and pervasive. If this is the case, the examples I gave earlier of inadequate knowledge of social and political issues and of history cannot be accidental. So it should come as no surprise that the majority of college students who have taken a basic physics course can still think, for example, that heavier objects should fall faster.[14]

Lingua Franca?

I noted earlier that I was particularly puzzled by the failure of most of my students to develop a sufficiently strong sense of the English language. Problems with grammar and syntax could perhaps be attributed to insufficient practice and feedback from professors. Inadequate pronunciation, though, is a lot more difficult to explain in this way. In addition to attending classes, reading, and writing, my students now spend a lot of time watching English-language movies and other programs. Yet, even some of the brightest among them now find it difficult to master basic (even if a bit crude) English pronunciation, and particularly to suppress the different pronunciation of Latin-derived words in their native language. At recent defenses of senior theses in our department, two Bulgarian students repeatedly mispronounced some very common words like "ultimate" and "public" (the "u" in related words in Bulgarian are pronounced as in "book"), "migrants" and "migration" (the "i" being pronounced as in "mill"), and a few other words. These were excellent students who should have heard each of the words they mispronounced multiple times, yet these had failed to sink in. I have heard students mispronounce even everyday words like "download" and "campus" which should have become etched a lot deeper into their brains.

I wish there were a more innocuous explanation for such linguistic glitches. I am afraid they do reflect, like learning problems in general, disruptions in brain plasticity and implicit learning. As increasing numbers of young people around the globe try to master English, such problems will likely contribute to the consolidation of a cruder form of global *lingua franca* (or "Globish"[15]). This, in itself, is likely to have an adverse effect on the way experts in different areas process textual information. There is some research indicating that reading a text in a foreign language evokes a blander empathetic resonance, a tendency which results in cruder utilitarian thinking.[16] If this is the case, then the "weird" Western mindset described by a now famous trio of psychologists[17] may eventually become universalized, at least among highly educated professionals and scholars.

These problems do predate the internet. As I already noted, Jane Healy catalogued related cognitive dysfunctions in the late 1980s on the basis of many interviews with teachers, as well as direct observations and studies. She directed particularly harsh words at "Sesame Street," the popular children's program which had failed in its ostensible mission to help generations of American children become better readers.[19] "Private Universe," the short documentary which exposed the shocking scientific illiteracy of students graduating with flying colors from Harvard, appeared in 1988.[18] A few years earlier, the President's report, *A Nation at Risk,* had already described the state of American public education as disastrous. And Rudolf Flesch's controversial book, *Why Johnny Can't Read,* had already identified a severe reading problem among American schoolchildren back in the mid-1950s.[20]

Flesch's book pointed very early to the deleterious effect of television, a concern which later became fairly widespread. His immediate intention, however, was to promote the use of phonics as the most efficient methodology to teach reading. Most subsequent hand-wringing over problems in education has similarly focused on perceived problems in the functioning and pedagogical underpinnings of the American school system and college education. This narrow preoccupation is best illustrated by the dramatic diagnosis offered by the authors of the "Nation at Risk" report. In a sentence that went viral long before the term was invented, they concluded that "if an unfriendly foreign power had attempted to impose on America the mediocre educational performance that exists today, we might well have viewed it as an act of war."[21]

What goes on in and outside the classroom is undoubtedly important. Still, I tend to think that a narrow focus on dubious pedagogical standards and practices, or on the time students waste exposed to screen-based media, can obscure deeper problems. I have come to believe that these problems reflect a more general crisis in learning which has only become more intractable with the IT revolution that truly took off in the 1990s. In this sense, the "crisis in education" Arendt once lamented may be broader and deeper than even she realized.

At the neurophysiological level, these broad capacities seem to depend on dopamine-mediated signaling. Some research findings suggest that declarative memory, implicit learning, sustained attention, and decision making are affected by levels of and sensitivity to dopamine in different parts of the brain. These need to be moderate and finely modulated, with fitting spurts and dips, and finely tuned responses to these.[21] Chronically elevated levels of and dulled sensitivity to dopamine, reflecting chronic overstimulation, are particularly counterproductive. These can lead to boredom and apathy, distractibility, and impaired learning. From this point of view, the crisis in student learning is primarily a crisis of warped brain plasticity.

The Unbearable Lightness of Being Understimulated

In a recent study, college students were placed in an empty room without any digital gadgets or other distractions, other than a device allowing them to self-administer mild electric shocks. They were to spend between 6 and 15 minutes alone with their thoughts. The majority did not really relish the experience. Sixty-seven percent of the male students and 25 percent of the young women chose to give themselves one or more electric jolts rather than complete the "thinking period" undisturbed. Before the experiment, most had indicated that they would rather pay five dollars than suffer such an electric shock.[23] No wonder that the study of "boredom" has become a hot academic topic.[24]

Our children's brains and our own do change in order to adapt to the increased abundance of various temptations and the ubiquity of screen-based virtual interaction in our lives. I suspect, though, that this kind of adaptation is detrimental to the ability and desire of students at various ages to succeed in formal schooling; and—more generally—to develop their thinking (and underlying neural networks) through reading both in and out of school. As *Washington Post* columnist Robert Samuelson has concluded in exasperation, the "meager results" from waves of educational reforms in the United States cannot be attributed mostly to commonly cited factors like inadequate facilities and resources, or the insufficient number, qualifications, or remuneration of teachers. In fact, there have been marked improvements in all these areas, with no noticeable results. Samuelson has therefore pointed to a "larger cause of failure [that] is almost unmentionable: shrunken student motivation."[25]

This troubling diagnosis in fact implies something that can hardly be stated—that American students (like those in many other countries) may have become less impressionable and by extension, "teachable." If this is the case, then the educational "attainment" demonstrated by students in a country like Finland may not result primarily from better pedagogical practices or teacher selection, training, motivation and the like. Finnish students might, for complex reasons, be more open to the kind of learning schools aim to foster, or generally more "educatable."[26] Unfortunately, the number of countries where the majority of students appear to benefit from such an overall learning aptitude is relatively small. In fact, there are now growing concerns about a global crisis in student learning which can no longer be attributed primarily to lack of access to education or of resourses.[27]

From such a perspective, trends in student learning would not appear as a direct function of pedagogical practices. Such shifts would be more strongly affected by the broader sociotechnological environment in which students grow up and develop. In some East Asian and European countries which have done well in the PISA studies (Shanghai and Hong Kong in China, South

Korea, Singapore, Finland, Estonia, and Poland) most children start formal education later—usually the year they turn 7. This is obviously not a sufficient condition for superior educational achievement, but may help. According to David Whitebread, a Cambridge University expert in cognitive development, the preponderance of empirical evidence shows "that children who have a longer period of play-based early childhood education, that goes on to age six or seven, finish up with a whole range of clear advantages."[28]

Whether excessive exposure to TV, the internet, or other visual distractions and social trivia can be classified as a full-blown "addiction" in the clinical sense is beside the point. What really matters is whether such massive daily exposure has a profound impact on brain wiring and neural processing. On the basis of the latest neuroscientific research, it seems there can be little doubt about this. And even a loose parallel with recognized chemical or behavioral addictions would suggest that the main effects of such a virtual immersion would include overall affective desensitization to weaker stimuli, and increased impulsivity.

Again, this is probably a vicious cycle in which greater exposure to virtual experiences begets decreased pleasure from and concentration for sustained reading—so only increased and more relentless virtual immersion can provide the level of stimulation which no reading material (or even "real" leisure activities) can offer any longer. These dynamics would be played out most forcefully in the brains and bodies of children and adolescents since these have, for better or worse, the strongest plasticity. As Carr has warned, however, even mature brains are far from safe[29]—and the digital onslaught or indulgence we all face is part of a much larger existential overload.

11

The Pursuit of Overstimulation

I already praised Marshall McLuhan for his most famous cryptic pronouncement—that "the medium is the message." He intended this aphorism to convey the idea that the nature of different information technologies affects the sensibility and nervous system of recipients more profoundly than the content of messages carried over various communication channels. Armed with this insight, McLuhan emphasized the mesmerizing effects of television and the way it was transposing the children who were growing up with it into a never-never land detached from the world of adults.

Such an understanding of the effects of that new medium could lead to the conclusion that television in and of itself provided stimulation that was much more compelling than anything children and adolescents had experienced earlier. TV screens can, indeed, be seen as "supernormal stimuli" which evoke a much more powerful physiological response than any natural objects or events.[1] In recent decades, these screens have become bigger and brighter, and the intensity of the stimulation they provide has grown. This effect is only intensified by the shift to high-definition broadcasts and TV sets[2] as well as the addition of 3D, internet connectivity, app support, and other add-ons to newer models. As communications professor Byron Reeves has noted, a big-screen TV "may turn up the volume on whatever emotional responses would have been experienced with a standard presentation."[3] Since focusing on something requires that the brain automatically turn down its reaction to everything else, lasting supernormal stimulation can make sustained attention toward what matters much harder.[4] Computer and gadget screens are smaller, yet interaction with them is a lot more intimate. They mediate over extended periods of time forms of social stimulation and access to curious trivia, often preoccupying every spare minute in their owners' hectic schedules. Moreover, use of digital devices tends to supplement extensive exposure to television, with the two media creating an increasingly compelling virtual cocoon.

According to neuroscientist Gregory Berns, what the brain really "needs" most may be information, pure and simple. In his words, "neurons really exist to process information... If you want to anthropomorphize neurons, you can say that they are happiest when they are processing information."[5] And, in pursuing this kind of info-grazing "happiness," we may indeed come to "resemble nothing so much as those legendary lab rats that endlessly pressed a lever to give themselves a little electrical jolt to the brain."[6] Like them, we are sensitive not only to the substances or experiences providing stimulation, but also to the cues we have learned to associate with these—and the latter can be dissociated from the former, and come to provide heightened sensory and affective stimulation in and of themselves.[7]

Once staring at glittering screens, particularly those providing access to the internet and various forms of virtual interaction, turns into a compulsion, affected individuals would need ever more intense and prolonged exposure to the constantly refreshed images, sensations, and information generated by screens of various sizes. They would seek to experience the previous excitement—but would be unable to get a lasting satisfaction, with the "real world" feeling slightly underwhelming against the backdrop of compulsive self-stimulation.[8]

A few years ago, psychologist Philip Zimbardo described in a controversial TED talk a more sinister version of this curse. His impassioned argument was followed by a TED e-book and a paperback (both co-authored with Nikita Coulombe) expanding on his initial points. [9] Zimbardo believes too many boys and young men succumb to a broadly defined "arousal addiction." They become hooked on video games and online pornography, and are in constant need of novelty and variety in these forms of gratification. Moreover, the hyperarousal associated with video games and online pornography is the tip of a larger iceberg of pervasive stimulation and virtual indulgence—all made possible by our enchantment with digital technology. Once young brains learn to crave and expect such overstimulation, the ability of their owners to feel similar excitement from books, ideas, and even physical objects (or human bodies) would tend to fade.

But even the combined power of television and screen-based digital attractions may not in itself be sufficient as an explanation for their hold over masses of "users," particularly children, adolescents, and young adults. Interacting with screens of various forms and sizes may become even more appealing for individuals whose nervous system has additionally been "softened up" by the compulsive consumption of a deluge of "supernormal stimuli." Neuropsychologist Deirdre Barrett, who popularized this term, has pointed to super tasty fast food, television, and pornography as some of the most obvious examples for such excessive levels of stimulation.[10]

The syndrome she describes, however, may be even broader. As Peter Whybrow has pointed out, our brains are evolutionarily adapted not just to weaker stimuli, but also to general scarcity. They are therefore poorly equipped to deal with the overabundance of goods and experiences churned out by a technologically advanced market economy. These can hijack and throw out of balance the dopamine system of the human brain and induce a state of "mania," or a permanently altered state of consciousness. This overall effect is reinforced as exposure to, and consumption of, "superpalatable" foods engineered to provide "supernormal" temptation and facilitate addiction,[11] novel "cool" products, excessive and often unpredictable social rewards, and so on.[12]

This overall picture should also include the pervasive,

incessant generation of enticing sensory cues through marketing and advertising. All these inputs, plus a deluge of screen-mediated essential and trivial information, almost constant access to sensory and social self-stimulation, other forms of real or virtual thrill seeking (including video games with in-built addictive properties), and overall stress form a powerful torrent of unrelenting overstimulation of the human brain's motivation system.[13]

Psychologist Fred Previc has similarly described a mass "hyperdopaminergic syndrome" linked to chronic overstimulation of the human brain's dopamine pathways and related epigenetic adaptations. In his view, such neurophysiological adaptations to the onslaught of modern living underlie more specific symptoms like emotional cooling and detachment, individualism, relentless goal-directedness, excessive risk-taking, utilitarian decision-making, multiple chemical and behavioral addictions, as well as the increase in recognized psychiatric conditions like autism, schizophrenia, hyperactivity, obsessive-compulsive disorder, bipolar disorder, mania, delusions of grandeur, etc.[14]

Previc's concerns have in fact received support from many recent studies which have contradicted earlier, more optimistic conclusions. Journalist Tony Dokoupil has reviewed some of these less sanguine findings indicating that our increasing digital immersion may lead to mental illness and clinical delusion.[15] He has noted wryly that "the black crow is back on the wire." He has also warned his readers not to kid themselves since "the gap between an 'Internet addict' and John Q. Public is thin to nonexistent." Previc, however, believes that even within the "normal" psychiatric range manic or hyperdopaminergic tendencies may lead to "delusional behavior and, in a milder sense, rationalization and denial."[16] Like Whybrow, he sees diagnosed mental illness as the tip of a much larger iceberg—posing more obvious problems on a spectrum of developmental skewing.

Unreality?

Previc's concerns are indirectly corroborated by studies that are different from those reviewed by Dokoupil. For example, survey data show that a growing proportion of American adolescents classified as "overweight" do not perceive their weight as abnormal.[17] According to other surveys, the vast majority of Millennials are optimistic they will be as or more prosperous than their parents—despite the hardships they face and dire predictions from experts.[18] In a 2011 Gallup poll, 64 percent of respondents identified "big government" as the biggest threat to the United States, with "big business" coming a distant second at 26 percent.[19]

Such findings do seem to capture some larger tendencies—which are corroborated by more impressionistic observations. For example, one can wonder at the overall persistence of the "American dream," or the idea "that the greatest economic rewards rightly go to society's most hard-working and deserving members," amidst fast-rising inequality (to levels antedating the Great Depression) and the progressive hollowing out of the middle class.[20] The fusion of reality show and politics in the bizarre persona of Donald Trump, and the grassroots resonance of the image he projects, seem similarly puzzling. Moreover, his main Republican opponent was someone who claimed to stand for America's humiliated and insulted—while he himself had been educated at Princeton and Harvard, had held senior government positions before he was elected to the Senate, and is married to a Goldman Sachs executive. The list could go on and on—including, of course, the continued flooding of the United States with military-grade assault rifles amid signs of growing mental disturbances among young males and a diffuse terrorist threat.

I do fear the changes in neural functioning induced by chronic stimulation can be analogous to those associated with recognized forms of addiction to chemical substances.[21] They can also be related, however, to an overall dependence on overstimulation (or self-stimulation) of the motivation system of the human brain—as opposed to any specific substance abuse or specific compulsive behaviors. Part of this broader syndrome is what psychologist Stephanie Brown has described as an addiction to speed and ever more hectic modes of work and lifestyles.[22] Such a "meta-addiction" is at the heart of consumerism and various other forms of continual sensation-seeking. Over the past two decades or so, this compulsion has been compounded by an intensified drive to induce new, potentially limitless and largely "virtual" needs and desires in millions of future consumers and money-makers from a very young age.[23] With the explosive growth of information technology, these tendencies have been supplemented by an ever intensifying "infomania."

The fatal attraction of a broad variety of "infomanic" behaviors, however, raises an obvious question. Why have so many of us, and of our children, succumbed to these if the dysregulation of the nervous system they induce has obvious existential downsides? As in the case of more specific addictions, focusing on the inherent addictiveness of chemical substances or compulsive behaviors may be misleading. It is widely recognized that many individuals who sample such substances and behaviors do not develop a full-blown addiction.

In the early 1970s, psychiatrist Lee Robins discovered that about 20 percent of American soldiers sent to Vietnam had become addicted to heroin. The majority, however, were able to kick the habit on their own after they returned home. Apparently, the high levels of stress in a combat environment had facilitated the development of what had been considered a purely chemical addiction; and once "addicts" returned to civilian life, most no longer needed the substance they had abused. Only a minority of former soldiers, those who perhaps had a more compromised nervous system as a result of previous experiences or had suffered more traumatic insults, and had found adaptation to "normal" life too overwhelming, relapsed.[24]

The role of stress in facilitating substance abuse has also been demonstrated by—what else—clever experiments with animals. As it turned out, the tendency of the proverbial lab rats to self-administer drugs to the point of complete exhaustion and even death was not entirely natural. In the late 1970s, Canadian psychologist Bruce Alexander set out to test this hypothesis by building what he and his colleagues called "Rat Park"—a giant enclosure offering a colony of lab animals a habitat as similar as possible to their natural environment. It provided sufficient space for free movement, exploration, socializing, and mating, and contained plenty of food and "playground" equipment.

Placed in this environment, most rats had very little appetite for sipping morphine-laced water as lab animals had done when kept in isolation in small, bare cells. Alexander concluded that under "normal" circumstances most animals and humans would not develop an addiction to the substances they would abuse under conditions of severe stress and deprivation.[25] But does human life in contemporary society resemble rodent life in Alexander's Rat Park? And could most individuals inhabiting such increasingly stressful and overstimulating social milieus ignore potentially addictive substances if offered virtually unlimited access to these? Or curb compulsive behaviors if these are spurred by powerful technological and socioeconomic forces? The obesity epidemic spreading around the world since the 1980s, which straddles the divide between substance dependence and compulsive behaviors, does not inspire much confidence in such human temperance[26]—and it is only one poignant example.

Of course, even many traditional communities have failed to develop the habitual sobriety Alexander might have predicted for them. They have long used mind-altering substances as a gateway to an altered state of consciousness and communion with the spirits inhabiting their worlds. Still, modern living seems to have induced a level of overall stress that produces a much more acute need for self-medication and stronger behavioral compulsions—a tendency which has markedly accelerated in recent decades. Such generalized distress is related to the overall increase in social complexity, pace of life, density of interactions, material abundance, technological saturation, and information overload associated with "late modernity." It is made even less bearable by the growing fragmentation and "disenchantment of the world," the loss of communal support and of spiritual horizons which can help most individuals withstand and make sense of the existential maelstrom engulfing them on a daily basis.

The suspicion that modern societies and technological progress create a deeply unnatural and potentially unhealthy habitat[27] is hardly new. It was shared not only by Romantic and conservative writers and thinkers. Nietzsche was particularly wary of the neural overstimulation individuals seemed to experience in the hectic, noisy, and overcrowded cities of the time. He worried about the "massive influx of impressions" that "press so overpoweringly—'balled up into hideous clumps'—in the youthful soul; that it can save itself only by taking recourse in premeditated stupidity." [28]

Many other intellectuals shared Nietzsche's testy premonitions as they observed the increased levels of audiovisual pollution and overcrowding in the growing industrial cities of the 19th century. In medicine, the increasingly frequent diagnosis of "neurasthenia" (or "nervous exhaustion") reflected similar concerns. The resonance of those worries, however, gradually weakened as the hectic bustle of the big city became the new normal. When in the 1920s German sociologist Georg Simmel raised similar concerns, he was a lot more sanguine about the trends he analyzed.[29] He described how urban inhabitants needed to develop an emotional distance from their daily experiences and from fellow city dwellers. In his view, such a "blasé" attitude allowed them to keep their psychological balance and function relatively undisturbed in a variety of social roles involving multiple social interactions.

In the 1960s, Simmel's more benign interpretation was questioned by Marshall McLuhan who described a degree of "auto-amputation" as a necessary adaptation to information overload.[30] In the 1970s, futurologist Alvin Toffler raised a slightly different concern. He argued that an unprecedented degree of sensory and cognitive overstimulation was causing a "future shock"—a state of mental confusion and "blurring of the line between illusion and reality."[31]

These impressionistic arguments received empirical support from a series of shrewd experiments designed in the 1970s by social psychologist Stanley Milgram. He observed patterns of communication among city dwellers, their willingness to help strangers, and other social behaviors. He concluded that the swarming of masses of diverse inhabitants in large modern cities created a psychological overload. As a result, individuals placed in such urban environments would be less likely to show compassion and extend help to others. In the scientific jargon he used, he defined overload as "a system's inability to process inputs"—either because there are too many, or they come too fast (or maybe both).[32] After Milgram, many other sociologists adopted the view that urban settings produce an information overload which could account for much of the malaise of modern living. Yet, they rarely put into serious doubt the ability of most individuals to strategically adapt to a potentially unhealthy urban environment.

With hindsight, the apprehensions of Nietzsche and other modernization skeptics seem far more prescient. I am often reminded of Henry David Thoreau who sought refuge in nature from Concord, which must have been a rather tranquil village in Massachusetts. Or Emily Dickinson who imprisoned herself in her family's estate. Or Marcel Proust who a bit later sought to protect himself from the bustle and dust of Paris in a room insulated with cork paneling, and with windows covered by heavy curtains. Or Edvard Munch who gave expression to his own mental exhaustion in a series of paintings which came to be seen as the paradigmatic expression of modern existential angst. The extreme reactions of such intellectuals, writers, and artists can be easily dismissed as the overblown, deeply idiosyncratic antics of a few disturbed minds. Certainly, Proust, Munch, and others did succumb to severe mental and physical maladies, and many more showed various subclinical symptoms. I tend to see them, however, as the proverbial canaries in the pit whose heightened sensitivity made them more vulnerable to the generalized overload of modern civilization.

These stresses, meanwhile, had a much broader influence on intellectual and artistic life. They probably contributed to the erosion of formerly rigid authority structures, standards of proper behavior, and even artistic tastes. Conservative Spanish philosopher José Ortega y Gasset once described how in the late 19th century the young European intelligentsia quite suddenly lost their taste for representational art, harmonious music, rhyme, and the conventional narrative of the great novels.[33] With a nod to this shift in tastes and attitudes, a few decades later Virginia Woolf (who eventually drowned herself) famously quipped that "on or about December 1910 human character changed."[34] That change found its most visible expression in modernist art, music, literature, and architecture, which all sought to break free from social conventions and contexts.

New artistic and literary fads were closely related to various quirks in personality and behavior—for example, the rise and disappearance of the urban *flâneur*;[35] or unceasing attempts to shock the bourgeoisie not just artistically, but also through daily provocations—like taking a turtle or a lobster for a walk. Meanwhile, new "social sciences" like psychology, sociology, economics, and political science sought to mimic the scientific method developed in the natural sciences in search of similarly "objective" knowledge. Even philosophy was swept by a similar trend with the rise of "logical positivism" and "analytic philosophy." In the 1960s, such modernist pursuits were disrupted by the rise of "post-modernist" artistic and intellectual trends. Despite their cultivated diversity, those were united by a common rejection of the search for objective truths and deep essences, repudiation of artistic and social hierarchies, preference for random sampling from different styles and epochs, an overall sense of ironic detachment, and praise for social disinhibition and the subversion of arbitrary norms.[36]

These changes in modes of artistic expression and knowledge paradigms are often attributed to an evolution in styles or ideas; or to the spread of particular new technologies like photography (which ostensibly prompted artists to leave behind attempts to represent aspects of "reality"), or the typewriter (which seemed to encourage a more telegraphic writing style). In my mind, changing artistic and intellectual fashions in the late 19th and early 20th centuries were more closely related to shifting sensibilities—reflecting modifications in patterns of neurophysiological functioning under the influence of urbanization, industrialization, mass literacy, increased social complexity, technological change, etc. These trends have long been recognized as benchmarks of social "modernization."[37]

At the heart of these neurophysiological adaptations has arguably been a growing affective-visceral desensitization and dissociation (mostly sub-clinical) under the impact of sensory and social overstimulation. That was essentially the process observed by Simmel and Milgram among inhabitants of big cities. Further changes in sensitivity were also measured by the researchers who tested German children several decades ago and postulated the development of a "new brain." And in recent years neuroscientists have observed additional modifications in the brain wiring of "digital natives."[38]

It was Tocqueville, however, who saw it all coming long, long ago. Reflecting on the social and psychological changes associated with the waning of "aristocratic society," he noted that "the bond of human affection is extended, but it is relaxed."[39] As a result, we would sympathize with the victims of an earthquake in Haiti, a deadly epidemic in western Africa, ceaseless fighting in the Middle East, risky voyages in the Mediterranean or Aegean, or vicious terrorist attacks. We may even make a modest or, in the case of someone like Bill Gates or Ted Turner, a sizable financial contribution to help alleviate acute problems in distant lands or closer to home. But we may not be moved too deeply by the personal misfortune of a neighbor or of a cousin living on the other side of town.

While Tocqueville admired many aspects of the "democratic" society he found in the United States, and had little doubt that it was the wave of the future for France and other countries, he was also suspicious of the potentially excessive individualism it fostered. He thought such individualism was qualitatively different from the selfishness which had existed even within tightly knit medieval communities. While that older selfishness was an expression of self-love, a strong emotion, individualism was colder and more calculating. As the new kind of society that had developed in the United States was based on civil equality and market exchange, it induced in individuals a false sense of self-reliance. In Tocqueville's view, such a social context could "make every man forget his ancestors" since it "hides his descendants and separates his contemporaries from him; it throws him back forever upon himself alone and threatens in the end to confine him entirely within the solitude of his own heart."[40]

This radical social detachment has also included the loss of the larger ontological or existential horizon associated with transcendent religious belief. In this broader sense, it was later noted and theorized by many other intellectuals who gave it different names: alienation, estrangement, disenchantment, desacralization, profanization, etc.; or wrote about an overall social malaise, anomie, ennui, civilization's "discontents," a loss of purpose and meaning by a "homeless" modern mind, etc. These concepts have slightly different connotations, but they all point to an existential void which opens as modern individuals become detached from previously secure communal or spiritual moorings.

This sort of disconnect is highlighted memorably in Leonardo DiCaprio's documentary, *The 11th Hour*. In it, a string of respected biologists and other scientists ponder essentially the same question: If we humans are so obviously part of nature, why have we forgotten this fundamental truth and adopted a blindly exploitative (and potentially self-destructive) attitude toward the natural world? Ironically, the answer to this all-important question may be partly rooted in our own biology. Onetime internet proselytizer Douglas Rushkoff has similarly lamented the disconnect he sees as pervading all of American (and not just American) life, attributing it mostly to a conquest by an omnivorous "corporatism."[41] Taken to an extreme, such tendencies can produce a nagging sense of dissociation and "unreality." Such sensations may be weaker than those resulting from acute trauma or overwhelming stress,[42] or may not even present themselves as abnormal or problematic. They can still be seen, however, as a diffuse syndrome—which can probably explain why the first *Matrix* movie struck such a chord.

At the heart of such existential detachment probably lies a common neuropsychological syndrome—a gradual numbing of the senses and gut feelings as a necessary adaptation to a more cacophonous, hectic, and technologically saturated social environment. Modernization has, indeed, often been associated with a degree of emotional cooling—a notion captured perhaps most strikingly by Marshall McLuhan's observation that "the Westerner appears to people of ear culture to be a very cold fish indeed."[43] Meanwhile, neuroscientists have come to recognize that the meaning or significance of particular experiences, events, and narratives is largely derived from their affective resonance and the general coupling of thought with physiological arousal.[44] But the link between affective-visceral desensitization and the loss of transcendence and meaning in the modern age has not been commonly made.

I do suspect the link between our overall desensitization and the loss of existential moorings (or fetters, in the more optimistic interpretation) reflects a general state of "meta-addiction" and pursuit of overstimulation—resulting in chronic existential (in psychiatric jargon, "allostatic") overload. This syndrome has developed under the influence of a social and technological milieu that has become increasingly overwhelming for human brains and organisms that are evolutionarily adapted to life in a mostly natural environment marked by general scarcity and face-to-face interactions within small communities.

Such a jaded sensibility is exacerbated by the ubiquity of screen-mediated accessing of information, communication, and gaming. It probably underlies most expressions of indifference or ironic detachment, desperate attempts to recapture a slipping sense of vitality and authenticity in contemporary societies, and increased apathy among students at various levels. This set of challenges inspired Bill McKibben to call at the dawn of the new millennium for a new kind of "mental environmentalism"[45]—reminiscent of Neil Postman's earlier notion of "media ecology."[46]

12

...and Existential Disconnect

As I was fleshing out this dispiriting diagnosis, I was beginning to worry that I had dug myself into a conceptual hole from which I could hardly climb out. To cheer myself up, I kept thinking about the minority of students in my classes who were still performing at a seemingly high cognitive level. How could their mental sharpness be explained? Were their peculiar aptitudes the result merely of superior natural ability which no amount of overstimulation could short-circuit? But what would then account for some anomalies in their thinking, particularly the more pragmatic and utilitarian focus they often displayed—to the point of expressing incredulity that an intelligent person could still opt for a non-utilitarian decision in moral dilemmas like the famous "trolley problem"? Or for the grammatical and syntactic problems still marking much of their writing? Moreover, there was also the famous "Flynn effect" to contend with—the alleged gains in "intelligence" of each successive generation (as indicated by standard IQ tests, if by little else).[1]

A plausible answer to all these questions came in the form of another epiphany. It was triggered by a brief report introducing yet another neuroimaging study which arrived in my inbox in October 2012.[2] I could have glossed over it and tried to file it away mentally or electronically. But as I was reading the summary of the experiments directed by neuroscientist Anthony Jack and some of his subsequent comments, the pieces of my mental puzzle started to rearrange into a new "neuroepistemological" framework. The latter could hopefully explain the cognitive divergence I observed as well as the ever increasing predominance of "positivism" (or "scientism") in the social sciences and policy research.[3]

Jack's experiments fell within a line of research that had shifted focus from the activation of different brain centers to that of widely distributed networks and their interactions. He focused on the interactions between two major networks. One was the "default mode network" which had already attracted growing attention. It had received its name from studies in the 1990s which indicated that it was activated "by default"—during periods of rest, when experimental "subjects" were lying in the gut of the scanning machine with no cognitive tasks to perform. On the basis of such observations, the initial consensus among researchers was that this was a "task-negative" network involved mostly in internal neural processing. Its spontaneous, rather high activation level seemed to underlie the generation of overall self-awareness as it integrated signals coming from subcortical emotional centers, internal organs ("interoception") or parts of the body (somatosensory and sensorimotor representations). At the conscious level, the default mode network thus appeared to be involved primarily in self-referential thinking and biographical recollections or projections.

The other network in the focus of Jack's study is commonly designated by an even more technical term—the "task-positive network." In previous experiments, it had been consistently activated when participants were asked to perform demanding mental tasks. It was therefore thought to be involved mostly in cognitive control, the logical processing of information, planning, and decision-making. It was also recruited, however, for tasks requiring attentional control. In addition to the volitional focusing of attention, it was activated by strong sensory (mostly visual) stimuli, whether those required immediate attention or needed to be ignored. In recognition of this role in the algorithmic processing of sensory or more abstract information and in attentional control, this network is sometimes dubbed the "executive attention," "executive control," or "central-executive" network. In fact, it largely overlaps with one of the two networks I described earlier with reference to seeking behaviors and addiction, the "reflective" system, and with the "executive brain" more generally.

In this context, it was the default mode network whose workings seemed more intriguing. First, it is involved in a strange dance with the task-positive network. In technical terms, the activation of the two networks is "negatively correlated"—when one "lights up" on brain scans, activity levels in key centers of the other typically fall below baseline levels. Second, some neuroscientists have noticed that the default mode network is activated not just during periods of idleness and mind-wandering. Apparently, it is also recruited when individuals need to distinguish animate from inanimate objects, infer the motivations and intentions of others, assess their own relatedness to different individuals and groups, and keep track of everyone and of social situations, interactions, and hierarchies.[4]

With hindsight, this larger role of the default mode network should hardly have been surprising. It was already understood that social judgment requires the empathetic simulation of the mental and emotional states of others—a process built upon somatosensory and sensorimotor representations and affective responses in the brain.[5] It did take time, though, for neuroscientists to begin to view and study the default mode network in this new light. What emerged from this line of research was a recognition of the substantial overlap between the default mode network and the "social brain"—the set of regions involved in the mapping and navigation of social relationships.[6]

This is where Jack and his team stepped in with their series of ingenious experiments. Those indicated that the default mode network was, indeed, involved not just in internal, self-referential processing, but also in a particular category of tasks requiring an outward focus of attention. The researchers asked participants to assess purely physical and social interactions—for example, whether water would flow from one container to another through a tube if there was a hole in it; or whether a young man thought the young woman sitting next to him (shown in a video clip) was angry. As expected, scans indicated that the physical puzzles activated the task-positive network in the brains of participants. Social scenarios, on the other hand, reliably activated the default mode network. In Jack's interpretation, those results offered a critical demonstration that the supposedly "task-negative," inwardly focused default mode network is also involved in the empathetic, affectively colored processing of social information.

Previous studies had already found that reasoning about physical objects and social judgment was associated with different brain regions. Jack's study helped clarify this picture by establishing that an outward focus of attention did not necessarily quiet the default mode network. It also indicated that the two modes of reasoning were largely incompatible since they required the recruitment of either the task-positive or default mode networks, whose activation alternated in a see-sawing fashion. One mode of thinking is analytic-empirical, and is appropriate for the causal explanation, utilitarian assessment, and prediction of physical phenomena and interactions. The other is empathetic, geared toward the partly intuitive, holistic understanding of social situations and interpersonal or communal give and take. The two modes do not normally mix. It seems the human brain has evolved to automatically switch between these two mental orientations or faculties—one social, the other non-social, physical or mechanical—as appropriate to any given situation.[7]

Since social and mechanical reasoning are associated with what had been called the "default mode" and "task-positive" networks, Jack has given these more evocative labels. He has called the former the "empathetic," and to the latter—the "analytic" network. These designations do seem preferable to the more technical terms they replace, but I would suggest a different term for Jack's "empathetic" network. As he himself has pointed out, the default mode network is involved not just in the empathetic understanding of other human beings and social relationships. It is also recruited for "some more *synthetic* forms of non-social reasoning, such as insight problem solving and detecting broader patterns."[8] Its proper activation is thus needed not only for social connection, navigation and positioning. It is also key to the holistic grasp of complex, indeterminate phenomena ("seeing the forest"), making distant associations between seemingly unrelated issues ("connecting the dots"), creative insight, an intuitive sense of what is relevant and significant, etc.[9] Crucially, Jack's "empathetic" network is also involved in implicit learning—acquiring a sense of probabilistic relationships and patterns (such as those involved in complex syntax and grammar) without conscious effort.[10]

Taking into account this broader role of the default mode network in our existential involvement and ontological grounding in the larger world, I would rather refer to it as the "empathetico-intuitive," or—for shorter—"intuitive" network. This designation still resonates with the "social brain hypothesis" according to which "our brains have expanded so much over the course of evolution precisely because of the challenges involved in living in large social groups."[11] Evidently, navigating this intricate existential web has required the development of broader neurophysiological and mental capacities which facilitate dealing not just with social interactions but also with complexity and ambiguity in general.

It remains a bit unclear what processes prompt the timely toggling between the analytic and intuitive networks and related mental modes. Some neuroscientists have assigned this role to a third, related brain network—the so called "salience network." It is involved in the ongoing monitoring of internal and external neural signals, on this basis perhaps prompting either the analytic or the intuitive network to be activated, and the other one to be inhibited. In any case, the determination of which network is to be recruited at any given moment seems to depend primarily on whether the task at hand or the context require keen affective and visceral attunement, or whether such internal processing would interfere with the successful handling of external information or with abstracted logical analysis—as when excessive ruminations could distract us from seeing an approaching car, solving a hard logical problem, or following an algorithmic protocol. This is, at least, how an optimally functioning brain, well integrated internally and with the body, has evolved to work. As in the case of addiction, though, this intricate balance can be easily disrupted.

Such neurosomatic unbalancing is most obvious in some psychiatric patients. On one extreme, individuals on the autism spectrum tend to have an underactive and underconnected intuitive network which is not coherently activated at rest, and is not easily recruited in social situations. At the same time, they have an overactive analytic network which is very hard to disengage—a peculiarity reflecting a general tendency for some neural deficits to be accompanied by related overcapacity in other areas. In such individuals, the more difficult activation of the intuitive network, and the simultaneous disengagement of its analytic counterpart, may also be related to insufficient sensitivity in parts of the salience network to internally generated neural signals or to their inadequate connectivity to other brain regions. On the other extreme, individuals diagnosed with schizophrenia and other psychotic disorders typically have a hyperactive and overconnected intuitive network which can hardly be quieted down.[12]

In both cases, the brains of affected individuals cannot switch smoothly between the analytic and intuitive networks and the related mental orientations. Those on the autism spectrum commonly remain stuck in a mode of external monitoring (often fixating on insignificant details), and of mechanistic analysis even in social contexts calling for empathetic sensitivity and a more intuitive judgment. A famous case in point is the 18th-century savant who was taken to see a Shakespeare play. Asked what he thought of it, he responded that the actors had uttered 12,445 words and the dancers had performed 5,202 steps—and he got the numbers right.[13] Individuals suffering from schizophrenia, on the other hand, tend to remain trapped in their inner mental world. They experience fanciful representations projected within their own minds as external voices, consider their vivid hallucinations truthful, overinterpret trivial actions and words, and often develop paranoid delusions (like mathematician John Nash's conviction that individuals wearing red neckties were part of a sinister communist conspiracy). Paradoxically, in both groups a failure to properly activate one network and suppress the other is socially debilitating as it entails a loss of touch even with significant others, and with the broader social "reality."

Of course, these are clinical extremes. But there is no sharp line dividing such extremes from the mental functioning of "neurotypicals."[14] There are, for example, many software engineers and artists who appear at least quasi-autistic or quasi-psychotic,[15] or in possession of "high-functioning" versions of the two potentially crippling conditions. Moreover, courtesy of brain plasticity, intense and prolonged engagement in tasks requiring different kinds of attention and mental processing can sharpen the contrasting tendencies and aptitudes of the two groups, eventually pushing some individuals over the edge. This danger is illustrated by the social ineptitude of someone like Christopher Langan (the person with the highest IQ score in history) who was never able to hold down a job where he could employ his exceptional mental abilities,[16] or the descent into apparent madness or self-destruction of many strikingly talented artists (and some mathematicians like Nash[17]).[18]

Madness or weaker psychotic tendencies among all sorts of creative types have always provoked strong interest. But I have come to see as more relevant and indicative the peculiarities of individuals who have been diagnosed with Asperger's[19] (a high-functioning form of autism), have remained undiagnosed though perhaps meriting a diagnosis, or display weaker, subclinical symptoms. I have kept wondering if such quirks might somehow be related to the somewhat unbalanced cognitive strengths of some of my better students, and also of most social scientists and all sorts of analysts or "knowledge workers" employed outside of academia.

Some of Jack's conclusions and comments may point in this direction. In his view, not only psychiatric patients but also "normal" individuals could depend too much on the activation of either the analytic or "empathetic" (as he prefers to call it) network.[20] But it seems the odds of leaning toward either extreme are not equal. As Jack has noted, formal education is generally aimed at tuning up the analytic network[21]—a tendency which becomes, I would add, stronger the higher one climbs up the education ladder. My sense is that not only education and any other form of abstract, logical processing, but also the degree of sensory overstimulation I described earlier and an overall increase in the cognitive demands generated by modern society could have a similar effect.

This impact first became obvious in Britain where industrialization prompted not only Romantic musings, but also the quasi-autistic, obsessive theorizing and plans for social improvement championed by Jeremy Bentham and his disciples; and where Adam Smith's more nuanced vision of society and the market were later replaced by cruder economic and social models. Those typically justified poverty and exploitation as providing necessary incentives for employment under the inhuman working conditions of early factories—or rationalized the extreme inequality generated by the "free market" as a necessary corollary to social progress through the "survival of the fittest."[22]

We have come a long way ever since—to a point where we inhabit a sociotechnological pressure cooker which requires and fosters some fairly unnatural aptitudes. These include the navigation of multiple information streams; abstracted, utilitarian analysis; and rapid switching of attention, ongoing choosing and decision-making in the face of countless options. It is these "analytical skills" that are valued, selected for, and reinforced as hallmarks of mental fitness in a modern, technologically supercharged social environment[23]—at the expense of a keener sense of the overall significance of changing social practices and broader trends, and of non-utilitarian or non-monetary valuations.

There was, however, one obvious problem with this theory as it was forming in my mind. If the sensory and cognitive workout provided by modern civilization had such a brain-building and mind-sharpening effect, why weren't the majority of my students displaying heightened levels of such mental fitness? Why did they tend to separate into a minority of exceptional (and often utilitarian) "learners" and a majority whose thinking displayed varying degrees of confusion,[24] detachment, and difficulty storing relevant information in long-term memory—and recalling it as needed? Why had the once solid group in the middle largely dissipated, as I and many of my colleagues had frequently complained?

As it turned out, these questions also had a plausible answer. It emerged in my mind as I stumbled upon an article co-authored by Mary Helen Immordino-Yang, a neuropsychologist who has done much research in the area of "neuroeducation."[25] The article highlighted the importance of the proper connectivity and activation of the "default mode network" for various mental competencies, overall human development, and education. It also contained, however, a stark warning—that excessive demands on the attention of children and adolescents in and outside of school could require overengagement of the task-positive (or analytic) network, and thus sabotage the development of the default mode (intuitive) network in their brains. Since the latter appears to form the "structural core" of the neocortex[26] whose proper development and attunement is essential for coherent engagement with the larger world, its inadequate activation and connectivity can leave students ensnared in concrete thinking and immediate associations. They would be unable to generalize across experiences and information streams. They would also have difficulty seeing the bigger picture into which their personal lives are inserted since they would lack a rich overall framework—whose development depends on ongoing implicit learning and the ability to make rich associations.

Another article by Immordino-Yang and several colleagues also suggested a link between this newer line of research focusing on the analytic and intuitive networks and previous studies of the localization of different functions in the two cerebral hemispheres of the human brain.[27] Generally, the left hemisphere is described as the seat of focused, sequential, algorithmic, logical-analytic processing. It is more self-contained and tends to generate more detached, forward-looking and optimistic plans or rationalizations for social behaviors. The right hemisphere, on the other hand, is associated with more synthetic and holistic representations. It is more strongly activated at rest[28] and more closely connected to subcortical parts of the brain involved in affective responses and interoception. It contains hubs of the intuitive and salience networks which are key to empathetic attunement, the integration of neural processing throughout the brain, and the overall integration of the brain and body.[29]

Curiously, the proper functioning of the left hemisphere depends on signals coming from the right hemisphere with its stronger sub-cortical connections. In cases of right hemisphere damage, affected individuals often succumb to false rationalizations, egocentric calculations, and even delusions; and more subtle forms of right-hemisphere malfunction can result in subclinical forms of delusional thinking.[30] These distortions are apparently produced by inadequate integration of affective and visceral signals into higher cognitive processes.

The separation of functions between the two hemispheres of the human brain and between the analytic and intuitive network thus seems to partly overlap and be mutually reinforcing. In the vocabulary of neuroscientists, the regions comprising the analytic network are "slightly left lateralized," and those of the intuitive network are "slightly right lateralized." This difference may turn out to underlie much of the differentiation in neural functioning and representation between the two hemispheres.[31] It could also provide a counterpoint to the usual dismissal of the contrast between "left-brained" and "right-brained" thinking as a myth.[32] Neuroscientists often point out that the two hemispheres are well connected, and are harnessed in tandem to perform various tasks. This observation would carry less weight if the activation of some key hubs in each is negatively correlated.

So here is the (provisionally) final picture which has emerged from all the cross-references and associations I have made so far. It seems the chronic overengagement of the analytic network (triggered by formal education from an early age, sustained cognitive effort, and our fast-changing sociotechnological environment) and the corresponding withering of the intuitive network tend to produce different "learning outcomes" at a deeper neurosomatic level. On one extreme, some students do develop strong "analytical skills," and an ability to keep track of and process (though not always integrate) vast information currents through sophisticated abstract frameworks.[33] Meanwhile, many remain caught in mostly concrete thinking detached from any larger frame of reference, some falling short of high school or college-level requirements. Such students tend to focus on issues and interactions deemed significant within their own "social networks" of similarly oriented peers (occasionally glancing at Facebook or other websites during class and school-related work outside of class). They remain understandably detached from the constellation of larger social issues I—and academics like Bauerlein and Edmundson—still find inherently significant.

There is also another twist to this story. The intuitive network is recruited not only for the purpose of social judgment or the holistic grasp of complex patterns. It is also involved in metaphorical thinking and, even more importantly, in implicit learning—the unintentional, unconscious absorption of information or knowledge.[34] This is the form of learning that underlies the acquisition of physical but also of mental skills and aptitudes. It is involved particularly in the mastering of complex rules and patterns (like those underlying grammar and syntax), but also seeing the proverbial "big picture" and developing the integrated understanding of the larger world that is ostensibly at the heart of liberal education.

Learning and automatically applying such skills and rules can, however, be disrupted by sustained mental effort and focused attention which recruit the analytical network in the brain. Sensory overstimulation and constant access to screen-mediated information has a similar effect. Downtime—made ever more elusive by the ubiquity of digital devices[35]—is essential for recovery from the "executive fatigue" induced in such an environment, and for the proper development and maintenance of the intuitive network in the brain.

Almost constant recruitment of the analytical network in the brain may result in inadequate activation and connectivity of the intuitive network. This neurophysiological unbalancing could in fact be the key to understanding the difficulties students in my classes, even many of the stronger ones, have with English grammar and syntax, with class material that appears overly "theoretical" or otherwise removed from their immediate frame of reference, and with integrating what they have learned in different courses into a coherent mental framework to which they can assimilate new knowledge. Curiously, such difficulties seem to go hand in hand with either overly literal or rigorously logical thinking—and lack of critical distance and sufficient self-awareness.[36]

Initially, the two groups of students in my classes—those deemed excellent by most academic standards and the many falling behind—appeared to have contrasting inclinations and capacities. As I kept thinking, though, the two mental toolboxes started to appear in a new light. They seemed to have something in common, related to the "weird" mindset described by psychologists Joseph Henrich, Steven Heine, and Ara Norenzayan. In a much discussed paper, they had criticized the tendency of their discipline to extrapolate from experiments with "weird" subjects (socialized within Western, Educated, Industrialized, Rich, and Democratic societies)—and to posit the highly unusual traits displayed by such exceptional individuals as psychological universals, common to the whole tapestry of humanity.[37] The hallmarks of this "weird" mindset appear to be excessive individualism and disproportionate faith in personal agency.

The "weird" thesis recalls earlier observations by cultural psychologists who have questioned psychological universalism. They have described two very different worldviews—one typical of Western societies, the other most pronounced in Eastern Asia but probably common to most non-Western regions[38] (with some quasi-Westernized areas and social groups around the world perhaps occupying a middle ground). The Western outlook has commonly been described as more analytic, narrowly focused, egocentric, and utilitarian—generally overvaluing individual traits, preferences, and actions at the expense of broader social forces and influences. The non-Western perspective, on the other hand, appears to be more holistic, communal, contextual, and existentially grounded.

Cultural psychologists have traced the divergence between these opposing worldviews to antiquity, linking them to differences in agricultural practices (focusing primarily on the communal effort and extensive infrastructure needed for rice cultivation in China), social and political organization (competition among small city-state vs. centralized imperial administration), institutionalization of intellectual pursuits (multiple private "academies" vs. state-controlled intellectual pursuits), etc.[39] Such cultural disparities may reflect, however, not just lessons learned and transmitted within societies with different cultural practices and norms. They could also be related to differences at the level of neurosomatic engagement with the world and overall existential grounding (or disconnect).

This link is demonstrated by studies indicating that cultural differences start at the deepest, least conscious level of neural processing. For example, when shown a picture of an aquarium, American undergraduates will typically focus on and recall details about the central object or the biggest fish. Their Chinese and Japanese counterparts, on the other hand, will take in the whole picture, and be able to describe the background more fully (while recalling fewer details about the main object). Also, Far Eastern experimental "subjects" tend to show activation in overlapping brain areas when thinking of themselves and their mothers; while in Americans these tasks trigger different patterns of neural activation.[40]

Within Western societies, pre-existing cultural and psychological tendencies have been reinforced by centuries of modernization. This is a complex process which is still not well understood, but has involved the gradual submission of social life to the logic of the market economy (what economic historian Karl Polanyi once dubbed "the Great Transformation"[41]), industrialization, urbanization, the weakening of communal bonds, expansion of formal education, growing technological saturation, various forms of overconsumption, sensory and social overstimulation, etc. The sum total of these trends has created more abstracted social relations, overall rationalization and institutionalization of social life, increased social density and complexity, accelerated pace of life, and the general sensory, social, and information overload I already described.

Needless to say, these tendencies have been taken to a whole new level with the development and proliferation of information technology over the past six decades—a sensory revolution which started with the spread of TV in the 1950s, and has been mightily accelerated by the information revolution. All these changes have imposed unprecedented demands on attention, and have required relentless analytic processing and a stream of minor or more consequential choices. Excessive exposure to screens may have a particularly insidious effect. They not only attract our attention, engaging the analytic network. They may additionally suppress activity in the intuitive network in the brain by inducing us to blink less frequently.[42] All these influences could play a major role in the kind of affective and visceral desensitization I described earlier.

All this seems like a recipe for further skewing of the balance between analytic and intuitive thinking, and the related networks in the human brain. This slant away from keen affective and visceral attunement and holistic thinking could account for the changes I have observed on a smaller scale among my students—particularly the divergence between a minority of "analysts" (in the broad sense) with overdeveloped "nerdy" cognitive powers; and a larger mass whose thinking is similarly narrow and utilitarian, but detached from any complex conceptual framework.

The first group seems very similar to the "empirical kids"[43] graduating from prestigious American universities (or at the top of their class in other institutions). The second group is more diverse and in some sense similar to their less stellar American counterparts—though not as self-absorbed and individualistic[44] in their thinking and ambitions. Nevertheless, even the academically weaker students across cultures are likely to appear more "intelligent" than their parents and grandparents if intelligence is reduced to a narrow set of cognitive skills. They would do better on IQ tests (particularly with the more extensive drill in test taking and algorithmic thinking they have received in school)—contributing to the rise in IQ scores psychologists interpret as evidence for increased intelligence, or the "Flynn effect."[45] While these two groups seem to have contrasting cognitive aptitudes, there is one quality they appear to share—the excessive disconnect from the communal settings and any larger existential horizon which Tocqueville feared individualism could eventually bring about.

As I already noted, this growing human estrangement was much resented by the Romantics. Later, it was captured by the string of philosophical, sociological, and psychiatric concepts I mentioned pointing to an overall existential estrangement and anomie. Of course, the sociotechnological trends I have described have affected non-Western societies as well, so the predispositions of children and adolescents there may be getting closer to the "weird" norm. Also, a dwindling number of individuals even in Western societies have been able to maintain a keener sensibility and a more holistic existential orientation (sometimes shading into mysticism or a "fundamentalist" yearning for moral purification). They, however, have often struggled and been marginalized within a social milieu to which their frame of mind and neurophysiological proclivities are not well adjusted[46]—a trend epitomized by the increasing marginalization of the traditional humanities in higher education.

Though their article is called "The Weirdest People in the World?", Henrich, Heine and Norenzayan have understandably tried to understate the "weirdness" of the disconnected mindset they have described. They have made a point of emphasizing that they did not intend to present any culturally shaped set of psychological traits as superior or inferior—just to suggest that academic psychologists should expand the pool of the test subjects they use to make it more representative of humanity. Science writer Ethan Watters, however, has read a disturbing message between the lines of their original paper. He thinks it very much reconfirms findings from earlier psychological research indicating that the more analytic "Western mind is the most self-aggrandizing and egotistical on the planet: we are more likely to promote ourselves as individuals versus advancing as a group."[47] For the sake of fairness, Watters could have noted that this mindset also entails toleration for different beliefs, values, lifestyles, and sexual preferences. He has pointed out, though, that individualist bias—at the expense of a more holistic grasp and communal belonging—has also constrained the thinking of mainstream Western social scientists.

In fact, Western social and neuroscientists must have an even sharper analytic focus—which can perhaps make it difficult to grasp the bigger picture and understand larger social patterns, trends and influences within their own and other societies. Needless to say, they tend to consider their own predispositions as "normal," and to see fanaticism and intolerance as aberrations in need of explanation (and debunking—usually in the form of academic high-minded conspiracy theories[48]). This outlook has also allowed most Western social scientists to maintain unwavering faith in a next generation of social "interventions" that will provide solutions to intractable social problems like poverty and large-scale violence.[49]

13

Unenlightened

Logically, all these problems could have a straightforward solution: curtail capitalism and runaway innovation, or perhaps even roll back technological progress. Or, as John Updike once pleaded with reference to the enhanced beer can that could be opened more easily, make sure progress has "an escape hatch."[1] Any such plans or hopes, though, are likely to remain a utopian—or perhaps dystopian—fantasy.

Meanwhile, there are well mobilized forces riding the wave of the Zeitgeist and pushing in the opposite direction. They are assaulting, among other things, what they see as the obviously dysfunctional American education system. In elementary and secondary education, there is the reform movement which advocates measurable benchmarks for student learning, teacher "accountability" linked to these, parental choice on the basis of such information, etc. This is a movement spearheaded by geeky entrepreneurs—mostly extreme "analysts" (in the broader sense) who have made a fortune in the IT sector. They tend to believe that the mechanisms which have worked so well in their companies could also be put to good use in education (and in any other area)—a conviction shared by economists developing mathematical models of education or related field experiments.

A related offensive unfolds both in elementary and secondary schools, and at the college level. It focuses on integrating technology in the education process[2] and preparing students for the "challenges of the 21st century"—rather than sticking to a model better suited to the bygone industrial age. The most radical approach in this area is to shift instruction increasingly online, with an emphasis on games and interactive applications which can keep students motivated and on task.[3] It has been almost two decades since journalist Todd Oppenheimer offered a spirited critique of this obsession with educational technology.[4] Yet it has continued to gain momentum, with some schools switching to digitized textbooks and even scrapping their physical libraries.

A more limited goal is to "flip" the classroom, or provide a model of "blended" learning—where students would preview learning content, and then use class time to discuss key points, solve problems, and complete various exercises (usually in small groups). An apparent advantage of this model is that a few star teachers could reach millions of students, while less stellar instructors would perhaps be more competent in a new role akin to that of teaching assistants. As an added benefit, students would stay engaged as they are asked to perform activities they find intrinsically rewarding.

In elementary and secondary education, the reformists' agenda and the shift to distance learning are supported by conservatives (and their well funded think tanks), since both offensives hold the promise of undercutting public education and the influence of the hated teacher unions. That overall reformist momentum will likely continue to gather speed as it is additionally spurred by private companies contracted to provide the learning "content" for various online platforms.

The emphasis on measurable outcomes and accountability has also been adopted by accrediting agencies at the college level. The more disruptive force there, however, is likely to be the spread of distance learning. Online or blended courses have kept some lesser campuses afloat, as they have been able to enroll larger numbers of non-traditional students. Such courses also allow colleges and universities to offer a more appealing (and probably cheaper) alternative to students who see traditional coursework as unnecessarily tedious and frustrating. "Edupreneurs" and a few prestigious universities have also embraced various forms of distance learning in an attempt to tap global demand. This shift is likely to put pressure on all colleges and universities to reconsider the way they package, sell, and dispense their degrees.

In addition to these more utilitarian and technological trends, there is yet another movement aimed at providing better pedagogical practices that could "reignite student learning" at different educational levels.[5] What do teachers need to do in order to accomplish this Herculean task? The answer is apparently simple: involve students at all levels in various forms of "experiential" learning and engaging activities like role-play, simulations, and hands-on projects (particularly ones that require collaboration in small groups). The idea is to reduce reliance on texts which growing numbers of students find unappealing and indigestible—offering instead activities and pedagogical interventions in and outside of the classroom that would be more stimulating and engaging.

In the minds of their proponents, all these approaches are bound to enhance learning and lift educational "attainment" at all levels. With my habitual skepticism, I have some serious doubts. I am tempted to retort that most education innovators are trying to make a virtue out of necessity, directing students away from learning activities with which most will struggle. Incidentally, such well-meaning reformers lead students away from the one practice which has historically fostered student learning and intellectual maturation—reading. And the new educational matrix offers a dubious alternative not just to students who would resist and be unable to master reading anyway. It could also discourage some potentially proficient, even avid, readers.

The importance of reading for brain development was once demonstrated by the eminent Soviet neuropsychologist Alexander Luria. He was a student of Lev Vygotsky, one of the early prophets of brain plasticity. During the Second World War, Luria studied the cognitive deficits and the physical impairments of Soviet soldiers who had suffered brain damage as a result of frontline wounds. He described memorably some of their symptoms, and devised exercises to help sufferers overcome or reduce their deficits. But before the war Luria had already conducted some research which, with hindsight, provided an equally powerful demonstration of neural plasticity, or the capacity of the brain to reorganize itself.

Can You Rebuild a Faulty Brain?

Luria's work had a tremendous influence on Canadian special education expert Barbara Arrowsmith-Young. In her childhood and youth, she had experienced severe cognitive deficiencies—for example, she could not tell the time when she looked at a clock, had to read a passage several times in order to decipher its meaning, and was physically clumsy. She nevertheless persevered and obtained a college education. Then she came upon Luria's work, and used it as a model and inspiration to develop a rigorous program that helped her overcome her specific neural deficiencies. She is now managing a special education school in which she is helping children with learning difficulties rewire their brains as she herself did.[6]

In the 1930s, Luria did field research in Central Asia aimed at studying the thinking patterns of local inhabitants. Those turned out to be unexpectedly—if stereotypically—"primitive." For example, when asked to name colors, illiterate herdsmen and their wives did not employ any abstract color names, but used only concrete terms associated with common objects, fruits, flowers, etc. They were similarly disinclined to give any abstract definitions of objects, or to group these into conceptual categories. They would resist putting objects (for example, a hammer and an axe) together as "tools," separate from other items on which those tools would normally be used (a nail or a log).

Most strikingly, the subjects Luria observed were quite unable (or unwilling) to formulate simple logical syllogisms in order to reach conclusions which surpassed their immediate experiences. For example, Luria and his assistants told respondents that all bears in the Far North, which is always covered by snow, are white. Then they asked what would be the color of a bear living on an island in the Far North which is always covered by snow. Most illiterate respondents would refuse to speculate on the attributes of an animal they had not directly observed.[7]

As Luria found out, only individuals who had received at least some formal education could engage in simple forms of conceptualization and logical reasoning. Those who had taken adult literacy courses were caught between the purely concrete thinking exhibited by illiterate nomads and a rudimentary degree of categorization and abstraction. Only the few who had acquired a more rigorous education in agricultural and other areas and worked on large-scale collectivized farms could usually achieve a degree of logical abstraction.

Luria's results became widely known only in the 1970s, and have since been replicated by anthropologists working in other parts of the world. They illustrate the way in which the brains of most people have worked throughout human history as they have adapted to the relatively simple cognitive tasks of life in traditional, technologically undeveloped, small-scale communities. Complex modern societies, however, require much stronger reasoning abilities and broader knowledge in many areas.

At the time, Luria did not realize clearly that he had observed brain plasticity in action. But the power of reading to reorganize the brain has now become widely recognized. One of its most prominent advocates is Maryanne Wolf, a researcher who studies reading and reading disabilities like dyslexia.[8] She has emphasized the extent to which reading is a deeply unnatural activity. In the brain, it requires the recruitment and complex coordination of many different areas which have developed through evolution to handle much simpler tasks. As a result of the sustained practice of reading, the brain therefore reorganizes itself—allowing individuals to become more capable of sophisticated thinking, emotional attunement, complex learning, and appropriate memorization.

All this could lead to the conclusion that I should advocate a form of cruel and unusual punishment: forcing children and college students to read, even if they do not wish to, in order to help their brains reorganize themselves the way Wolf posits they would. In fact, I am not so extremist or reactionary—and I doubt forced reading would work for most students if tried. Part of the reason for this is that different forms of reading are not created equal.

There is, in fact, one variety of reading which stands above all others. This is the immersive, "deep reading" to which I have already referred. It has been described perhaps most lucidly by English professor (and avid reader) Sven Birkerts.[9] It involves an absorbed, earnest engagement, in the very act of reading, with the written page and the imaginary worlds unfolding behind it. A form of "flow," or reading "in the zone," if you wish—or at least something close. "Deep reading" thus entails the kind of affective-visceral response that, according to neuroscientists, can trigger the release of brain chemicals facilitating the formation and consolidation of synaptic connections, and perhaps related epigenetic changes.[10] This form of reading also sends strong signals along nerve fibers connecting parts of the brain which are removed from each other, facilitating their myelination—and thus brain integration.

Such "deep reading" can in fact be a version of the "deep practice" journalist Daniel Coyle has described as essential to myelination. He has popularized (alongside fellow-author Malcolm Gladwell) the idea that around 10,000 hours of such "deep practice" are essential for mastery of any skill.[11] So maybe this is roughly the amount of reading students need to complete in order to become expert readers, with appropriately myelinated pathways in the brain.[12] This seems like a logical conclusion, but there is a catch. In order to have such a profound effect, reading does need to be "deep." Alas, this is easier said than done.

In fact, the general capacity for "deep reading" probably started to decline generations ago. A glimpse into this process is offered by German historian Rolf Engelsing who has described a profound transformation in the nature of reading. In his account, before the mid-18th century educated people practiced mostly "intensive" reading. They read repeatedly from just a few books, most notably the Bible. Their emotional engagement with those few readings was, in fact, very intense. Words and passages resonated deeply with their rhythm and the rich associations they evoked, and "became deeply impressed on [readers'] consciousness."[13] This process was probably assisted by the practice of reading aloud or subvocalization in reading to oneself. In the words of Birkerts, that was a form of "ferocious reading" which assumed or created unmistakable depth.[14]

In Engelsing's view, once the quantity of books, pamphlets, newspapers, magazines, brochures, and other printed matter increased, reading changed. The practice of "intensive" reading gave way to a more relaxed form of "extensive" reading. People started to read many books and periodicals, usually once and much more superficially. Though those sources lacked the same emotional resonance, many readers were still able to acquire a wealth of knowledge on various issues and develop increasingly sophisticated ideas.

The amount and variety of reading material did increase rapidly, but there was perhaps a deeper reason for the changing nature of reading. I suspect these changes were related to the overall shifts in sensibility in a modernizing society I described earlier. Once the number of readers increased and they started to read less intensely, the demand for new reading material exploded, deepening yet another feedback loop. Many decades later, we are now caught in a fast transition to a whole new form of reading. As we sample in quick succession scores of texts and images from the web, these evoke a feeble emotional response and leave little trace in the brain—except for the craving for even larger doses of digital stimulation, and a growing detachment from more serious texts and even the "real" world.

Nicholas Carr has become the highest-profile Cassandra sending repeated warnings that too much "infograzing" online could make us "stupid," or subvert the way our brains work.[15] Most provocatively, he has argued that we are fast developing a form of "artificial" or machine-like intelligence as our thought processes become detached from rich emotional responses. The transition to "extensive" reading Engelsing describes, however, indicates that fundamental shifts in reading and mental processing probably predated the tyranny of the screen, even the smaller black-and-white TV screen. Web evangelist Clay Shirky has pointed out that most readers had lost interest in *War and Peace* long before the Kindle edition came out—and for a good reason. In his view, "old literary habits," like reading a bulky novel from cover to cover, were an unfortunate "side-effect of living in an environment of impoverished access."[16] Apparently, fast and constant web access provides not just superior alternatives to immersive linear reading, but can also alleviate the guilt at shunning bulky classics.[17]

In less extreme forms, Shirky's optimism is shared by most educational experts who have sought to apply findings in neuroscience to education. Their advice is mostly aimed at increasing student enthusiasm and engagement in the classroom through the use of familiar or newer techniques—in order to "make learning NEW, EXCITING, and REWARDING."[18] This task can ostensibly be achieved through, for example, presenting material in ways that engage all the senses; using surprise and humor, or asking students to make predictions, in order to periodically jolt them into a joyfully attentive state; staging role-playing activities, preferably in period costumes; using "human interest" anecdotes; deploying different forms of experiential learning in the classroom;[19] giving examples or asking students questions which prompt them to link broader and more abstract topics to their personal lives; or using the capacity of the computer, particularly in gaming mode, to capture students' attention. The overall idea behind all these techniques is to get students into a "warmed-up state of alert stimulation" conducive to memory formation and consolidation. In the words of Judy Willis, a neurologist-turned-teacher, the use of exciting classroom techniques should help "keep this generation of students from falling into the abyss of joyless, factory-style education" through which she had to suffer.[20]

Most of this advice is targeted at elementary and secondary education, and seems to give a scientific grounding to older approaches seeking to encourage active, student-centered learning. Many of these approaches, however, have also been embraced by college professors, even by some who teach large lecture classes. Among the true believers, physics professor and YouTube sensation Walter Lewin has offered perhaps the most striking example. His lectures, before a packed auditorium, included many complexly choreographed, circus-grade stunts meant to illustrate basic physical principles. Despite his advanced age, he would cheerfully ride a fire-extinguisher-propelled tricycle across the stage in order to demonstrate jet propulsion; would stand still as a large metal ball suspended on a string swung and stopped an inch from his nose; or he himself would dangle, Miley-Cyrus-style, on a wire to demonstrate vividly the physics of pendulums.[21]

Since very few faculty members can match Prof. Lewin's commitment, resourceful inventiveness, and acrobatic (or theatrical) skills, many have instead tried to liven up their classrooms by inviting their students to become actors. In the humanities and social sciences, students have been asked to reenact various social and historical dramas—for example, dress up and behave like the Puritan settlers who once hunted alleged witches in their midst; or imagine they are representatives of different communities in a "world simulation" as they try to work out realistic solutions to global problems and humankind's contemporary predicament.[22] Meanwhile, many faculty members who have stuck closer to the traditional lecture format have sought to turn their classes, in the words of Mark Edmundson, into "laser-and-light shows, fast-moving productions that mime the colors and sound and above all the velocity of the laptop"[23] (and, one could add, of the gaming console).

In addition to stimulating classroom techniques and practices, higher education has been affected by a few additional trends. Students have been encouraged (and often given academic credit) to pursue various forms of experiential learning outside of the classroom—mostly through "service learning" and internships, but also through numerous extra-curricular activities which in some cases overshadow traditional academic learning. They have also been urged to spend a semester or year in a study-abroad program in order to develop their cultural sensitivity and empathy. All these practices are driven by the conviction that important—perhaps the most important—kinds of learning can take place outside the classroom and the traditional academic curriculum.

The belief that the traditional classroom belongs to the already distant 20th century offers inspiration to another trend in higher education which may have much more profound effects. This is the integration of online experiences into the educational process, to the point of offering online degrees. The most spirited defense

Extreme Education?

A few years ago, *Harvard Magazine* offered a sobering glimpse into the lives of its undergraduate students. Their overscheduled days—and nights—seem to be dedicated not so much to classes as to multiple extracurricular activities (in which, ideally, they would obtain a leadership role).[24] One student interviewed for the article represents a particularly disturbing case (at least for those who tend to cringe more readily). She "can describe different levels of exhaustion. One level, she explains, is a 'goofy feeling, like feeling drunk all the time; you're not quite sure what's going on. Then there's this extra level of exhaustion, where you feel dead behind your eyes. The last four weeks, that's where I've been. I get sick a lot.'" This, in fact, comes close to the delusional states induced by forced sleep deprivation or sensory overstimulation. Such overachievers typically say they love this lifestyle and cannot imagine a less intense routine.[25] Unfortunately, there is recent research involving "animal models" which suggests such chronic exhaustion and sleep deprivation can result in the withering of neurons in key brain centers.[26]

of such potential "distractions" is perhaps offered by Cathy Davidson who has drawn on her experience as a teacher and vice provost for interdisciplinary studies at Duke University. She has argued that the excessive focus at the heart of traditional pedagogical approaches in fact prevents professors and students from noticing some major issues.[27] The example proverbial she points to in the title of her book, *Now You See It*, refers to a now famous video of two teams of basketball players passing the ball. When asked to count the number of passes made by one of the teams, viewers typically fail to notice a student dressed as a large, hairy gorilla walking across the court.[28]

Davidson's argument is that the human brain is "built for distraction" by evolution, and to be human is to be distractible.[29] In her view, the demand for sustained focus can thus be dehumanizing. Once this unreasonable expectation is dropped, students are bound to thrive on the distractions offered by the internet, including sampling of information, blogging and sharing thoughts and content, the democratization of judgment through the crowdsourcing of grading, online gaming, and all sorts of multitasking which can be "multiinspiring."[30]

Even without such elaborate theoretical justification, many colleges and college teachers have already moved aggressively to incorporate online elements in traditional coursework (with the goal of "flipping" the classroom discussed earlier and freeing up time for more engaging activities). There are also, however, many courses and degrees offered completely online, and this may well be the wave of the future—despite some initial problems with student motivation and evaluation. A few years ago, MIT, Yale, Harvard, and a few other academic heavyweights started to provide free access to taped lectures and other course content without awarding any credit to students accessing these materials. Less prestigious schools (both in the United States and Britain) did not hesitate to make the next logical step and start offering full credit for online courses and even awarding online degrees.

This seemingly inexorable trend was later given a boost which may eventually take it out of the gravitational pull of traditional college education. The push came courtesy of a few high-tech startups like Coursera and Udacity which attracted venture capital with their promise to extend MOOCs (Massive Open Online Courses) to millions of eager students worldwide (an effort matched by the edX consortium involving Harvard, the MIT, and other top schools). Some of these companies have been started by university professors who have taught successful online courses to thousands of students and hope the new format will vastly extend their audience—and allow them to do well while doing good. Some distance-learning entrepreneurs, however, are just that—entrepreneurs, eager to "disrupt" yet another outdated industry.

In late 2013 one of the most high-profile players in the field, Udacity, stumbled and announced that it would shift its focus to more technical training—a change of direction which prompted many skeptical comments and predictions. It will be difficult, though, to put the genie back into the bottle or even scale it back substantially—particularly if new microgenerations of digitally attuned students become more accustomed to online interaction, and come to prefer it to live talk or exchange.

Champions of online education often cite as a successful example the now famous Khan Academy—a vast collection of short instructional videos. Initially, those covered mostly mathematical and natural science topics, but later their range was greatly expanded. Some high school teachers have started to ask students to watch the Khan video clips at home, using class time for problem solving and other forms of "active learning." If the new distance learning startups are successful, however, this model could infuse and perhaps "disrupt" higher education as well—to the benefit of a few brand-name universities, private companies, and star academics, at the expense of most other educational institutions and college professors (and of thousands of students who could still benefit from a more conventional curriculum—and personal touch).

14

Toward Edutopia?

When I want to relieve some of my anxieties, I often think not just of Yana (the student who mastered the English articles in one brief summer), but also of Sonya—another almost ideal student who graduated a few years ago. "Almost," because I would have loved to hear some of her thoughts in class. Alas, she never raised her hand to say anything—but her writing was superb. Her command of written English was so unusual that I thought she had perhaps lived and studied in an English-speaking country. She had not. When I asked her, it turned out Sonya had gone to a Bulgarian high school very similar to those attended by other Bulgarian (and a few international) students. Yet, she had somehow attained a degree of writing ability very few could achieve. She attributed her own success to having been an avid reader, mostly of fiction.

Perhaps part of the reason I find Sonya's example so inspiring is that it reflects my own experience. When I was fourteen I went to a Bulgarian high school in which the first year was dedicated mostly to the intensive study of English—the same kind of school Sonya had attended much later. Until that point, I had had zero exposure to English, and the first few tests I completed were pitiful. Later, I was gradually able to move on, acquire a sense of English grammar and syntax, and expand my vocabulary. During my undergraduate education in Bulgaria I continued to take English classes twice a week. Yet, throughout those 10 years in high school and college (interrupted by two years of conscript military service) I did very little writing in English. I was never explicitly taught how to write a paragraph, an essay, any other shorter format, or a research paper in English (or in Bulgarian, for that matter). And the English-based milieu of the internet had not yet materialized. Still, I kept reading widely in English, both fiction and non-fiction on various topics. Then, in my late 20s, I enrolled in the Ph.D. program in political science at Notre Dame.

By all accounts, I should have struggled there. In addition to my lack of training and experience in writing, I had had limited exposure to mainstream American social science. I was mentally preparing for the worst, but was pleasantly surprised. I was able to process the hundreds of pages of assigned readings every week, make sense of the main ideas, and expand rapidly my personal frame of reference. More importantly, it turned out I could write—though, as I already admitted, not as a master stylist. During my first semester, one of my professors (who taught ancient Greek philosophy) found it hard to believe no one was editing my papers. Another professor later noted my written English was more solid than that of most American students in the program. It took me six years to do all the coursework and research, and to finish my dissertation. This, however, is faster than the average in my field, and I was the first one among our cohort of 12 to defend my Ph.D. dissertation.

I wish I could say I was able to accomplish these feats because of some exceptional intellectual ability. My hypothesis, though, is more modest. I suspect I, like Sonya, benefited from being an avid reader from a fairly young age. Incidentally, this was not predetermined in any obvious way. My parents had grown up in a small village and obtained their higher education degrees (in accounting and mechanical engineering) in their 30s, as non-traditional students working full-time. We had only a few non-technical books at home. Moreover, I spent my first seven years in school in a class of 40 students (of whom only four, including myself, later went on to obtain higher education degrees).

Overall, my family and early educational background would probably be considered "disadvantaged" by American standards. But this did not hold me back. I made frequent trips to the public library in my small hometown, and to the home of an uncle and two older cousins who owned maybe 200 books. My parents were worried that there was no system to my reading. I, nevertheless, persisted. I would reread a favorite book like Raffaello Giovagnoli's *Spartacus* (a 19th century Romantic depiction of the Roman gladiators' revolt) six or seven times, and would frequently revisit favorite passages—a version of the "intensive reading" described by Engelsing (or the slightly weaker form of "deep reading" cherished by Birkerts). At the same time, I spent several hours every day playing freely in the street, as was still common in Bulgaria back in the 1970s—an aspect of childhood whose developmental value has recently received renewed appreciation by experts.[1]

I found the reading I did outside school assignments profoundly enjoyable—until in high school I reached a point where the experience could give me, quite literally, a high. Later, I experienced this neurosomatic effect even more reliably from reading non-fiction books and feature articles introducing exciting ideas. I sometimes mention this effect of reading half-jokingly to students and they laugh, but I am completely serious.[2] It is perhaps this personal experience of combining the joy of reading with educational "attainment" which has cemented my conviction that the road to meaningful reading and solid writing does not pass so much through various exercises in and outside the classroom. I suspect these can achieve only limited results without sustained, intent reading.[3]

This theory is hardly revolutionary. Even someone like cognitive psychologist Daniel Willingham (who places the usual focus on engaging classroom activities and pedagogical strategies) has concluded it is still essential for teachers to do whatever they can to "get children to read."[4] As psychologists Anne Cunningham and Keith Stanovich have noted, "early success at reading acquisition is one of the keys that unlocks a lifetime of reading habits"—which in their turn can help "to further develop reading comprehension ability in an interlocking positive feedback logic."[5] For better or worse, this is the only road to accumulating the background knowledge needed to process, make sense, and remember new information in any area and as part of an expanding mental matrix. Gaming, role-playing, debating, or other forms of experiential learning can never become such a constantly available tool of lifelong learning and knowledge accumulation.

As I mentioned, however, there is insufficient recognition in the pedagogical and neuroeducational literature that the "crisis in education" at all levels largely reflects a crisis of reading and related brain plasticity. Moreover, reading itself is often reduced to a technical "skill." There is, in fact, much empirical data to demonstrate that this "skill" depends critically on background knowledge, which also helps the absorption of new knowledge.[6] Unfortunately, the ability of students and adults to practice this skill and to expand their personal frame of reference (and to derive some satisfaction in the process) has suffered a marked decline.[7]

I have little doubt that this is, indeed, the heart of the educational conundrum we now face. And I think I know fairly well what it would ideally take to raise a generation of competent and enthusiastic readers. It should all start with healthy development in the womb, breastfeeding, and close bonding with parents.[8] To achieve this, parents should be a bit less overwhelmed by work, other responsibilities, multitasking, or stress and insecurity. As psychologist Catherine Steiner-Adair has pointed out, they first and foremost need to come back home unplugged from all their devices—and remain so for prolonged periods of time. This is the only way they can pay undivided attention and connect emotionally to their children.[9] It is hard to believe, but almost half of American infants may now be lacking such a vital, in the most literal sense of the word, emotional bond.[10]

Younger children in particular crave such connection, as if they sense how detrimental parental neglect and detachment can be to their development into individuals with appropriate emotional responses to the people and the world around them.[11] From birth, parents should engage in plenty of lively, warm communication with their children. This is something that cannot really be faked since even babies are especially sensitive to the loving attention of caregivers, to the tone of their voices, facial expressions, eye contact, etc.[12] When necessary, such parental involvement should be supplemented by forms of childcare which mimic it as closely as possible.

This kind of affectionate communication should not be interrupted frequently by screen-based activities and interactions. Crucially, children should not be given an electronic device (or placed in front of a screen) as a digital pacifier.[13] As I already noted, attachment to gadgets can partly become a surrogate for human relations;[14] and screens do not seem to evoke the same kind of emotional response as live communication. This is probably the reason why children cannot learn to speak by mostly watching TV or streamed video[15]; and why "Sesame Street" has not sparked the revolution in reading and learning its creators aimed for[16] (though the edutainment it provides has always attracted large audiences—a lasting success which will be more fully monetized on HBO). As psychologist Victoria Dunckley has noted, these problems may only be exacerbated by digital interactivity.[17]

Crucially, children should be regularly read to, and should hear plenty of exciting stories from a very early age.[18] Such stories typically contain more complex vocabulary and syntax than speech directed at children. Moreover, reading usually takes place "in the context of face-time, of skin-to-skin contact, of the hard-to-quantify but essential mix of security and comfort and ritual."[19] This is an experience most toddlers cherish, and they can listen repeatedly to the same age-appropriate tales of speaking animals and dragon-slaying or non-violent heroes. This kind of pre-reading is probably essential for helping most children achieve the cognitive and emotional (and, at rock bottom, neurophysiological) development which would prepare them to master and enjoy reading when the time comes. Regular free and pretend play with other children also seems essential for the ability of children to stay emotionally attuned to their social context and to modulate appropriately their own emotional responses.[20]

Once children enter school, the role of primary caregivers is largely taken up by teachers. From that point, the exact formula for success becomes much more elusive. Ideally, schooling should perhaps combine methods borrowed from Waldorf, Montessori, Reggio Emilia, and other similar approaches emphasizing "natural" learning and building upon the natural curiosity and developmental potential of children.[21] Formal education should probably wait until most children are developmentally prepared for it, maybe around the age of seven—as is done in Waldorf schools and some of the countries whose students excel in the PISA studies. Needless to say, this runs counter to the trend in the United States and many other countries to expose children to a structured curriculum emphasizing abstract mental operations (including reading, writing, and arithmetic) from an ever earlier age.[22]

Once started, schooling would need to be centered around the gradual development of reading and writing abilities. For better or worse, this is the only kind of training which can integrate different parts of the brain in a way that allows broader empathic attunement and engagement with complex texts, ideas, and larger social issues—providing a path toward open-ended, independent learning.[23] As Jane Healy argued at the dawn of mass personal computing, "the most complex neural systems, which pull together abstract language and visual reasoning, develop only if challenging encounters with reading, writing, and verbal reasoning continue during the teenage years."[24] This is an experience that should start in elementary school[25] and continue uninterrupted—unless at some point teenage rebellion or chronic dejection demonstrate conclusively its futility.[26]

Extensive cursive handwriting may be particularly important in this context since it seems to activate the brain in a more beneficial way as compared to print handwriting, and particularly to typing.[27] According to a much discussed recent study by psychologists Pam Mueller and Daniel Oppenheimer, students who take notes by hand have better recall and understanding of class material.[28] The authors attribute this result to the fact that longhand is slower than typing, so students need to digest and capture the essence of what they hear and see in class—rather than transcribe everything almost verbatim. It could also be the case, though, that the fine motor movements producing handwriting activate the sensorimotor cortex differently—enhancing learning.

Other opportunities to practice their fine motor movements, manipulate various objects, engage in artistic activities, simply play, or generally develop their "sensorimotor-perceptual skills" can provide younger students with additional healthy stimulation.[29] Such seemingly technical "skills" in fact reflect a process of brain maturation which is essential for higher-order cognitive functioning and engagement at later educational stages. Physical activity (not just sports or fitness, but also dancing) during school hours and adequate sleep can give a much needed boost to this process of overall neurophysiological development and maturation.

Do as I Say...

A much commented OECD study has found that increased use of information technology in the classroom does not improve the performance of students.[30] When required to study from e-texts, many students have complained of eyestrain—or what has been dubbed "computer vision syndrome." They have also showed greater distractibility and worse recall of the material. According to linguist Naomi Baron, such problems may be inherent in the medium itself. Digital texts seem to encourage interruptions and goal-oriented accessing of information (if not outright trivia seeking), or what Baron calls "reading on the prowl"—as opposed to more absorbed "deep reading."[31] About 90 percent of the students she surveyed said they preferred to read for courses from a hard copy. Still, the shift from reading in print to reading from screens is continuing, contributing to what Baron sees as "students' mounting rejection of long-form reading."[32] Meanwhile, many Silicon Valley "geeks" (some perhaps involved in the digital "disruption" of education) have opted to send their own children to Waldorf and other non-traditional schools that shun electronic devices.[33]

As far as the subject matter of schooling is concerned, it may require an approach different from the one underlying the Common Core curriculum now adopted by most American states, with its emphasis on decontextualized "skills." If students do not develop the ability and willingness to engage earnestly with assigned readings, the basic facts and ideas covered in any core curriculum will remain random bits and pieces detached from any broader framework. In most cases, such fragmentary information will not "stick" in the minds of students and they will not develop the ability to process new facts and concepts effectively and independently.[34]

To facilitate commitment to long-term memory, social and historical developments and personalities should be presented mostly in the form of vivid stories which are intriguing, even exciting. Such narratives should ideally be written by truly gifted writers. They will never be as riveting as the exploits of Harry Potter and his friends or of the young participants in "hunger games" and other thrilling exploits, or as enchanting as the *Twilight* series. Still, they need to come close, with elements of mystery, suspense, and intrigue. As neuroscientists Charan Ranganath has pointed out, "curiosity recruits the reward system, and ... seem[s] to put the brain in a state in which you are more likely to learn and retain information, even if that information is not of particular interest or importance."[35] Such evocative stories should also employ humor, and depict riveting examples of courage, devotion, accomplishment, character building, and justice served.

Dramatized in this way,[36] the knowledge schools and teachers seek to impart will have a stronger affective-visceral resonance—an aspect which is crucial for the ability of students to commit to long-term memory those major facts and ideas, and to grasp the overall significance of historical events and larger social trends (like, for example, the first anti-colonial rebellion in Haiti or the partial merger of personal liberty and social control in contemporary societies). A degree of ambiguity, or even confusion, may serve a similar purpose.[37] As neuroscientist Matthew Lieberman has argued, to be successful education must fully engage the "social brain"[38]—and this should not be limited to activities meant to foster experiential learning.

I am not sure to what extent this approach can apply in the teaching of math and science, other than adapt what educators like Bill Nye the Science Guy and actor Alan Alda have done outside the classroom.[39] In any case, there may be a need to rethink whether most middle and high school students should be expected to master highly abstract theories and mental operations. As neuroscientist Guy Claxton has quipped, "there is no good reason for inflicting trigonometry on everyone."[40]

Dramatized Learning

It is now common for students at all levels to be asked to act out historical or literary plots in class. I am not sure what this achieves other than making the classes less tedious and leaving student with some lively recollections of the performance itself. I believe empathic understanding, the expansion of personal mental horizons, and textually-scaffolded learning in general are better served by a level of dramatization in the text itself. My favorite example is a story Benjamin Barber uses to illustrate the threat consumerism may pose to human liberty—a hidden danger whose comprehension is linked to a peculiar understanding of "liberty" itself:

> In thinking about modernity and modern capitalism, Max Weber spoke a century ago about an iron cage. Consumerism brings to mind a different cage. There is a fiendishly simple method of trapping monkeys in Africa that suggests the paradoxes which confront liberty in this era of consumerism. A small box containing a large nut is affixed to a well-anchored post. The nut can be accessed only through a single, small hole in the box designed to accommodate an outstretched monkey's grasping paw. Easy to reach in, but when the monkey clasps the nut, impossible to get out. Of course, it is immediately evident to everyone (except the monkey) that all the monkey must do to free itself is let go of its prize. Clever hunters have discovered, however, that they can secure their prey hours or even days later because the monkey—driven by desire—will not release the nut, even until death. Is the monkey free or not?[41]

This story, whether true or fictional, never fails to trigger a lively classroom exchange. Dalrymple's mock defense of cannibalism[42] has worked similarly well—while also alerting students of the need to develop their sense of the language.

All these stories should be collected in physical books which are not overloaded with bulleted lists, tables, and charts—so they are not brimming with "visual noise." Such books should not be overly dry and schematic either, introducing instead any "content" in rich, metaphorical language. They should also be beautifully designed, impressively illustrated, printed on glossy paper, and bound in materials whose texture and feel enhances the overall substantive/tactile sensation. Such instructional "materials" would help provoke a stronger emotional response in young learners compared to the typical bland textbook bent on systematizing knowledge, to say nothing of laptop and tablet screens and "interactive" educational software (or "smart" boards and even desks in the classroom[43]).

The internet and screens cannot really be kept out of schools. But their use should never come to dominate the classroom and the out-of-class experiences of students at any level. Role-playing computer games, live simulations, and various forms of experiential learning should not crowd out reading and writing either. These are likely to produce stronger biographical memories of the activities involved as opposed to cumulative knowledge of significant events and ideas.[44] There is, in fact, some research which suggests that learning is more productive when it takes effort and deliberate concentration on sometimes unenjoyable tasks.[45] As psychiatrist Norman Doidge has concluded, "some teaching techniques abandoned back in the sixties as too rigid [including the dreaded rote memorization] may be worth bringing back"[46]—to the extent, perhaps, that these are still a realistic and beneficial option for most students. In any case, schools should ideally provide a temporary sanctuary from the pull of the internet, "social media," and gaming to which most students dedicate so much of their "free" time.[47]

Reading-centered learning should ideally start in elementary school, and continue through undergraduate education. In the process, the responsibility for engagement with increasingly complex texts should be shifted only gradually onto students. They could thus develop progressively their ability to understand and reason with increasingly abstract concepts. Along the way, they could build an expanding personal frame of reference to which they would be able to associate new information. This level of cognitive development is key for the "transfer" of knowledge—the ability of students to apply knowledge and broad principles acquired in one context to other contexts and issues (for example, apply Isaiah Berlin's "two concepts of liberty" to the way Hobbes, Burke, or Mill understood that central idea; or realize that contemporary forms of individualism and self-absorption can be seen as a realization of some of Tocqueville's worst fears).

Achieving such a level of conceptual thinking is not really possible without extensive, yet sufficiently intense and attentive reading. So it should come as little surprise that far too many students cannot, under current circumstances, reach sufficient educational maturity. Jean Piaget observed that in early adolescence children should normally reach a point when they can operate with abstract concepts.[48] This is the time when a final growth spurt, followed by the pruning of superfluous neural connections, helps establish a relatively stable architecture in the human brain. The latter then becomes the basis for further gradual neural modification and, in the best of circumstances, more complex and emotionally attuned thinking. It can be partly modified courtesy of brain plasticity, but can hardly be reshaped completely—and may be susceptible to many unhealthy influences later in life.

I fear the overall process of brain maturation which underlies productive social and cognitive engagement can be more easily derailed by various noxious influences than most experts allow. These are mostly related to sensory overstimulation[49]—which could even interfere with the pruning of inessential synapses and the neurophysiological optimization achieved through that process. Unhealthy influences also include, however, other forms of chronic overtaxing of attention (including visual overstimulation, frequent switching of attention and multitasking), as well as demand for repeated analytic operations from an early age.[50]

Among all these activities, the continuous excitement and overstimulation provided by moving images, virtual self-expression, accessing online social trivia, and various digital micro-pellets dispensed by websites, games, apps, or other programs are particularly hard to resist. Even online reading may not be very far removed in its effects on young brains. According to some researchers, reading from a screen is less cognitively demanding, and apparently evokes a weaker emotional response than reading from a physical page.[51] Apparently, digital reading also predisposes readers to focus on concrete details rather than get the overall gist of a text.[52] Unlike reading from a real page, online reading involves a lot of searching, clicking, scrolling, page loading, micro-decision-making (where to go or not go next), as well as less frequent blinking. These all contribute to reduced activation of the intuitive network in the brain which is involved in implicit learning, social and more holistic understanding, and a broader sense of existential grounding.

Deep, immersive reading from a physical page can in fact be inherently "interactive" if the reader is truly engaged with the text—in a way that calling up, viewing or scrolling down digital pages can hardly be. It yields stronger cognitive and emotional attunement, and deeper comprehension of complex texts and ideas. It also creates a mental representation involving the same neural networks that would be activated by similar situations in real life—and can thus facilitate the development of these networks, and of empathy.[53] For most readers, these effects are strongest when they read good fiction, and much weaker with formulaic fiction and most non-fiction. In any case, reading from a real page seems qualitatively different from "screading." The latter appears to be more distracted, casual, and aloof—a diminished form of reading which lulls the brain and can hardly produce intense affective engagement with the text.[54]

A Bibliophile's Confession

Writer William Giraldi provides a memorable description of his obsessive, unending love affair with books. He offers some memorable quotes emphasizing the physical aspects of such a relationship:

> There is no true love without some sensuality. One is not happy in books unless one loves to caress them.
> —Anatole France

> Thick tomes have traveled with me for thousands of kilometers across the face of Europe and have returned with their secrets unviolated.
> —Aldous Huxley

> The physical presence of this book, so substantial, so fresh, the edges so trim, the type so tasty, reawakens in me, like a Proustian talisman, the emotions I experienced when, in my youth, I ordered it.
> —John Updike (recalling a book on Japanese history)

Giraldi also refers to a striking confession by 19th-century physician and poet Oliver Wendell Holmes:

> I must have my literary harem ... where my favorites await my moments of leisure and pleasure,—my scarce and precious editions, my luxurious typographical masterpieces; my Delilahs that take my head in their lap...; the books I love because they are fair to look upon...

Giraldi believs this need for a sensual relationship with bound books can never be satisfied by a digital device—a conviction I fully share.[55]

As reading specialist Anne Mangen has pointed out, "materiality matters," and this goes well beyond the sense of touch.[56] This is a crucial point. It seems the human brain can, indeed, distinguish between the real and the virtual. Imaging studies have demonstrated that, for example, print ads tend to evoke a stronger affective response in the brain compared to their virtual counterparts.[57] When exposed to objects or products, and to images of those, the brain also tends to favor the "real thing." We may be tempted to think that if this is the case, maybe we can trust our brains to habitually prefer the real world to virtual reality, and unfailingly sense the difference. Unfortunately, the human brain can be easily duped. For example, the stronger affective response to a real object over its virtual reproduction is eliminated by simply putting the item behind glass.[58] More disturbingly, an untrained brain will almost invariably react with stronger excitation to the stream of electronic overstimulation which triggers the strongest release of dopamine—but provokes little emotional response to anything beyond the visual stream itself;

until even that excitement becomes blunted.

Old-fashioned reading may now seem superfluous as it is being replaced by new "literacies" and multitasking aptitudes better suited to the digital age. Incidentally, the development of the kind of abstract reasoning Piaget posited and traditional literacy once facilitated was not a strict necessity for most of the illiterate herders Luria once observed. This form of thinking, however, has become essential for our ability to function as competent, relatively adjusted members of the highly complex, informationally and technologically saturated societies of today. Unfortunately, spending half or more of their waking hours calling up and scrolling down mostly trivial bits of information and images may not help the majority of students climb over the developmental threshold once described by Piaget—particularly when practiced from a very young age. If the kind of brain maturation which underlies more sophisticated thinking is stalled during the adolescent years, it is much harder to achieve once students move to the post-secondary educational level. This may be the main reason accounting for the "limited learning" Arum and Roksa found across American college campuses,[59] and the fact that professors teaching upper-level courses cannot assume that most students bring along much accumulated knowledge and conceptual thinking.

Those who do scale the heights of abstract reasoning, on the other hand, face a different challenge. Their thinking can become overly detached and mechanical—and they would not even know it.[60] Incidentally, there is something the two groups share. They may both find it difficult to truly benefit from liberal arts education with its emphasis on developing some specialized knowledge within a sophisticated yet integrated perspective of the social and natural worlds.

The End of Analogy?

According to a study cited by Daniel Willingham, only 30 percent of American college students were able to see the analogy between two similar problems, each described in a single paragraph, which had an obviously similar "deep structure."[61] Another study has found that approximately half of graduating American high school seniors can be classified as "concrete active learners." They prefer direct, immediate, sensory information and experiences; have difficulties with reading, writing, and conceptualization; and are dependent on immediate mental gratification and step-by-step guidance from instructors.[62] Unsurprisingly, analogy questions which had long been part of the SAT exam generated many complaints—and in 2005 were removed (ostensibly to make room for an essay—which has now become optional).[63] As Arum's and Roksa's findings perhaps indicate, college students who experience such difficulties would struggle to develop a more sophisticated mode of thinking.[64]

Piaget is often maligned for his old-fashioned emphasis on abstract reasoning (which might have reflected the overly algorithmic nature of his own thinking processes). These criticisms may, indeed, have a point. Recent research in neuroscience has demonstrated that thinking needs to be infused with appropriate affective and visceral attunement through the proper connectivity and recruitment of the brain's intuitive network. Otherwise, individuals could hardly have sound judgment and make responsible choices in their daily lives and in positions of authority.[65] Both the minimum of conceptualization which is required for knowledge acquisition and transfer and the keen affective and visceral attunement needed for responsible social functioning require a certain neural substrate.

This neurophysiological foundation cannot be developed without absorbed reading; which, at a later stage, could beget more reading as increasingly sophisticated thinking and the background knowledge or frame of reference associated with it make new topics relevant and interesting.[66] As a beneficial side effect, such a virtuous circle could help students become more selective in accessing information and rationing interactions over the internet, and in their overall use of digital devices. They might then be able to set clear priorities and adopt the "information diet" David Shenk once recommended as a strategy for cutting through the thickening "data smog" which threatens to engulf us all. Before children and adolescents reach such maturity, it must fall upon parents and guardians to institute an "electronic fast" to help "reset" their charges' brains and partly restore their neurophysiological sensitivities.[67]

Meanwhile, the overall sidetracking of neurosomatic development I have described might explain why most high school and college students struggle with very similar cognitive difficulties; why observations made for one level of education can generally be applied to the other; and why the majority of college students apparently show either no or very limited gains in learning as they go through four years of "higher education."[68] If I wanted to be provocative, I could suggest a telling analogy. Teaching college students who have not made the age-appropriate cognitive leap Piaget described can sometimes feel similar to the experience of some anthropologists. These well-intentioned, committed scholars have sometimes spent years trying to teach members of indigenous communities which have not developed the concept of numbers (and have words in their language only for "one," "two," and "many") to count to 10. In some cases, they failed since their "students" lacked the mental (and neurophysiological) apparatus needed to engage in the required abstract operations.[69]

Even under the best of circumstances, some students will still fall behind, come to detest school,[70] seek electronic stimulation, and even be admired as "cool" by their peers who perceive their lack of interest in old-fashioned reading as a challenge to outdated norms and adult authority. For such students, or students with recognized learning disabilities, it may turn out that digital, screen-mediated and interactive content accompanied by exciting classroom activities provides an indispensible learning experience which can keep them engaged. But this kind of technologically upgraded education should not be forced upon students who have the potential and can develop the desire to become earnest and engaged readers.

15

The Last Oasis

At this point, I would likely be expected to offer a solution to the problems I have described. Both Whybrow and Previc, while warning about the dangers of chronic stimulation and dopamine overload in the brain, emphasize that the neurophysiological changes they describe can be reversed.[1] Of the two, Whybrow is notably less skeptical. He opens his latest book with an unsettling question he kept asking himself: "But why had the madness become so pervasive in Western culture?"[2] Then, in the second part of the book, he offers a comprehensive plan for taking personal responsibility, living well, building a "well-tuned" brain, and resolving the existential problems plaguing contemporary civilization. Whybrow has also dedicated a chapter to changes in education that could help diffuse the manic-addictive tendencies he describes.[3] He believes that, "ideally, formal education should be an extension of early parenting"—building upon ties of care and attachment, and tapping the "young students maturational drive to forge and hone the adaptive skills that will serve lifelong autonomy and self-directed growth."[4] Other authors have also felt a need to devise personal solutions to the similar problems they describe.[5]

Zimbardo and Coulombe have offered a more systematic approach. They hope the dangers they highlight can be addressed with the joint efforts of government agencies, schools, media and technology companies, parents, and young men and women made aware of the new challenges.[6] Even neuroscientist Susan Greenfield ends her latest book on "mind change," a far-reaching shift in the human mind and brain (analogous to "climate change"), with a four-step action plan.[7]

A Credentialed Digital Cassandra?

Professor Baroness Susan Greenfield has drawn the ire of many neuroscientists with her dire warnings regarding the overall impact of digital technology, particularly on young brains and minds. The musings of amateurs like Nicholas Carr (or myself) can be easily dismissed—not so those of a fellow brain "scientist." Or perhaps they can? The British Medical Journal has published an editorial faulting Greenfield for making baseless alarmist speculations in the media—rather than publishing her "claims in the peer reviewed scientific literature, where clinical researchers can check how well they are supported by evidence."[8] One of the editorial's authors, Vaughan Bell, had previously written a scathing review of Greenfield's book, *Mind Change.* [9] In it, he had accused Greenfield of systematic "confusion between correlation and causation and ... apparent inability to distinguish which sorts of studies can provide the best evidence for each." As a result, her whole argument had "the tone of a bad undergraduate essay in which a series of randomly encountered scientific findings are woven together with a narrative based on free association and a burning desire to be controversial." So Greenfield may be an esteemed neuroscientist, but in fact lacks the basic research skills expected from a decent undergraduate student, and is bent on alarmist sensationalism. Iain McGilchrist, a prominent psychiatrist and author, faced similarly dismissive rebuttals after he warned that many children were displaying borderline "autistic" behaviors as they spent so much time interacting with screens.[10]

My hunch is that such unguarded accusations reflect to a larger extent the predispositions of their authors rather than the qualities and qualifications of the person being castigated.[11] In her book, Greenfield deflects such criticism by claiming the right to draw broader conclusions from multiple studies and experiments. She also points to a common logical fallacy that should be familiar to her detractors (and perhaps even to undergraduate students): "absence of evidence is not evidence of absence."[12] The Baroness could have added that Cassandra's curse, after all, was to tell the truth. In any case, is this fundamental disagreement a matter of different competence or intelligence? More likely, it reflects disparate unconscious predispositions deriving from differences in affective-visceral processing, and the extent to which the latter is integrated into higher-order thinking.[13]

Many experts, meanwhile, do not seem to recognize that adaptation to a more demanding sociotechnological environment may have significant downsides—since there is no conclusive evidence for this. Neuroscientist Elkhonon Goldberg has even speculated about a political evolution analogous to the evolution of the brain. He has suggested that the world is undergoing a "transition from a world order build of a few large autonomous geopolitical units to a network of many small, highly interdependent geopolitical units."[14] In Goldberg's view, this network could come under some sort of global authority—mimicking the integration of the brain through signals from the frontal lobes.[15] Curiously, he kept that hope even following the September 11 attacks and the violent disintegration of Iraq.[16]

I do believe "we" are now facing an unprecedented existential crisis. Yet I would rather avoid the "last chapter problem" identified by professor of history and journalism David Greenberg. He has noted that almost every book analyzing social or political problems, "no matter how shrewd or rich its survey of the question at hand, finishes with an obligatory prescription that is utopian, banal, unhelpful or out of tune with the rest of the book."[17] I would rather not become one of the countless social critics who, in Greenberg's words, have "succumb[ed] to the hubristic idea that they can find new and unique ideas for solving intractable problems."[18] In fact, I suspect the solutions most authors have offered reflect mostly their own "positivity bias" and faith in human agency—tendencies that are strongest among the educated elite, particularly within "weird" cultures. Though these predispositions seem weaker in Greenfield, even she cannot quite escape the biocultural force field in which she is suspended.[19]

Free of any similar inclinations, I cannot in good faith offer a grand solution of my own. I suspect no amount of deliberate "interventions" can counter on a mass scale the perfect storm that engulfs us all. Alas, this storm seems set to gather strength given the competitive pressures of global capitalism, the relentless and explosive growth of information technology, the pull of the supernormal stimuli it generates, and the rapid proliferation of companies whose business plans are premised on subverting our neurophysiological balance and reflective self-control.[20] As already noted, the effects of all these noxious forces are be strongest on children and adolescents whose brains (and bodies) are still exceedingly plastic. So I am hardly surprised by evidence that programs aimed to boost executive control and non-academic abilities like "grit" in elementary school students have had little impact on overall academic achievement.[21] Or by studies which have found that a large proportion of high school students (or the majority in some high-pressure schools) display symptoms of mental illness.[22]

I believe, though, that we should look into the abyss and grasp the full extent of the challenges facing technologically advanced societies and the different generations inhabiting them. In my view, the trends I have described require a more evocative label, the way "global warming" is more resonant than the emotionally cooler term "climate change." We could refer to a "Learning Hijack Syndrome" resulting from chronic overstimulation and dopamine overload. Or, if a more technical term still seems preferable, perhaps "Hyperdopaminergic Learning Syndrome" would fit the bill (with a nod to Previc's notion of an overall "hyperdopaminergic syndrome" affecting late modern society[23]).

From this more holistic perspective, I have come to doubt that student success (or lack thereof) is primarily a function of pedagogy. I also suspect educational "attainment" is mostly undermined by the invasion of information technology and the introduction of more stimulating activities into the classroom—including digital self-stimulation by students accessing "social media," sending and receiving text messages and email, surfing the web, playing games, etc. with their smartphones, tablets, or laptops. In a recent survey, American college students admitted they were spending on average 20 percent of class time using digital devices for activities unrelated to class tasks or content.[24]

The goal of reading-centered learning and appropriate neuro-somatic maturation (to say nothing of any solution to the broader problems I have addressed) may already be out of reach at the societal level. The tide of the future is quite clearly flowing in the opposite direction, and I do not quite see how it can be stemmed or even weakened. It seems a lot stronger than anything any educational "interventions" can accomplish—no matter how much they struggle to make a virtue out of necessity. Moreover, many such schemes to enhance learning and build an ability to focus (on the basis either of neuroscientific research or of experiments seeking to establish practically "what works") may only reinforce the sensory overstimulation students face outside of school. As noted earlier, some "evidence-based" interventions and cutting-edge techniques and technologies might help the most disadvantaged or least motivated students, or those with learning disabilities. They could also benefit many budding "analysts" with various interests and career prospects. But such interventions could be detrimental to students who still have the potential to achieve deeper immersion in reading as the central learning activity, and to become keenly attuned to larger social issues.

The difficulties which make any hopes for a long delayed educational revolution unrealistic start from birth, if not earlier. From the womb, children are often exposed—and remain exposed—to an array of noxious influences. Too many are born by C-section[25] or are not breastfed (or are breastfed for only a short period).[26] A large number may not receive sufficient warm care and attention from parents who are overworked and frequently spaced-out—glued to their own digital devices and not fully "there" for their children. When this is the case, behavioral addictions can become a "poor substitute for love" and mutually attuned interactions.[27]

Part of truly "nuclear" families, similar partnerships, or single-parent households, children are typically deprived of extended contact with grandparents and other extended family members. They are also handed over to various ersatz forms of caregiving when both parents need (or prefer) to work for pay. And too often children come from an early age to rely on screens to provide stimulating interaction and entertainment. This cocktail of noxious influences has only thickened since former primary school teacher and literacy expert Sue Palmer sounded the alarm for a "toxic childhood" syndrome a decade ago.[28] These influences can be particularly detrimental to the highly sensitive children and adolescents described by psychologist Elaine Aron[29] and journalist David Dobbs, among others.[30]

Excessive screen time became an issue for students of all ages long before the arrival of broadband internet access and handheld digital devices. As I noted earlier, Marshall McLuhan believed that growing up in front of TV screens (like the protagonist of the 1990s sitcom *Dream On*) helped open the gap between the younger and older generations back in the 1960s. Then, in the 1980s and 90s, there was an apparent increase in the degree of understanding between parents and children who had grown up in a similar media environment. With the coming of age of the Millennials, and of each succeeding microgeneration immersed ever more intensely in a screen-mediated environment, some psychologists have described a new "generational divide in cognitive modes"[31]—even if many "digital natives" maintain regular, friendly communication with their parents.

Increasingly, schools at all levels are contributing to the deepening of these trends. Until recently, the school served to establish some balance in the lives of students by requiring them to work with the help of textbooks, general-purpose books, notebooks, and other "real" instructional materials. With the proliferation of faster and uninterrupted internet access, laptops and tablet computers handed out by schools, instructional video games, etc., the balance between screen time and all other activities has been tilted even further. Those countless hours spent staring into a screen entail a mostly sedentary, indoor-bound lifestyle, unhealthy snacking, insufficient sleep, and other unhealthy habits—a form of "co-morbidity" which cannot really be beneficial.

Once children are hooked on this constant stream of self-stimulation, books or even the "real world" are bound to hold a lot less attraction and to become a source mostly of boredom.[32] Even when forced to read from a real page (or, increasingly, from a screen), many adolescents would struggle to achieve the deep-reading immersion which is essential for learning and brain development. Alas, this is a syndrome which for the majority of students is likely to continue into college and beyond.

As Martha Herbert and others have argued, all these potentially harmful influences can add up, and trigger unhealthy neurophysiological adaptations.[33] In some cases, these adaptations will be expressed in the form of recognizable psychiatric symptoms, specific learning disabilities, or other "disorders." But they can also be diffuse, and associated with disruptions in learning and neurosomatic maturation even in the absence of a clear psychiatric diagnosis. If these difficulties prevent children from developing their ability and desire (or at least willingness) to read, more complex forms of thinking, and a capacity to navigate their dizzying social and technological environment, then the learning and overall developmental "outcomes" from education are likely to disappoint. Over a decade ago, a Canadian study found that less than half of the sixth- and ninth-graders who participated appeared developmentally mature for their age.[34]

Making these observations, I could easily be accused of repeating age-old complaints about the young, and blaming them for their perceived failings. As I look at the political and social clouds thickening around the globe, I keep wondering if the older concerns were entirely misplaced. To the extent they were, my fear is that this time it is, indeed, different—and we may be at a civilizational breaking point. A point where the long-term transformation sociologist Norbert Elias dubbed the "civilizing process" (the progressive internalization of appropriate social inhibitions accompanying the rise of the nation state and the market economy)[35] is increasingly unhinged at the neurophysiological level.[36]

Of course, these troubling trends will be shrugged off by observers who, like Virginia Heffernan, Hanna Rosin, or Steven Pinker, are not easily disturbed by any cultural permutations. If, however, the shifts in aptitudes and sensibilities I have described are to be taken seriously, children and young people can hardly be blamed for these. Young brains and bodies need to adapt to an environment that requires and fosters new skills at the expense of others. We would hardly blame penguins for losing their ability to fly while developing other capacities in order to survive in their harsh Antarctic habitat. Children and adolescents have found themselves in an analogous predicament affecting mostly their mental—but also physical—aptitudes.

If anyone carries any responsibility for the analogous adaptation I have described, it is "we," the older generations. We, in Bernard Stiegler's harsh judgment, have abandoned our obligation to take proper care of the young. The French philosopher, however, does not seem to accuse the older generations either. In his view, its members now inhabit "a society that has become *structurally* incapable of educating its children"—as it has been reshaped by larger socioeconomic and technological forces.[37]

This conclusion reflects a more holistic and—by extension—more fatalistic outlook.[38] It will ring hollow to readers and various experts with more analytic and optimistic predispositions. Unlike such critics, I have few illusions that the perfect storm of unhealthy influences and activities I have described could somehow be extinguished or curtailed. I do recognize that my own typically Bulgarian mindset may be overly apprehensive. But I suspect chronic optimism has its own pitfalls. Such an upbeat attitude can take you to the Moon and to victory in major conflicts like World War II and the Cold War. Incidentally, it can also get you into Vietnam, Iraq, and Afghanistan—or a spiral of economic and technological innovation with much unforeseen socioeconomic, political, and neurosomatic fallout. Indeed, Cassandra was right. As were the fear and moral panic mongers who later warned that Rome could be sacked.

In any case, the self-stimulation made possible by digital devices and screens of various sizes seems just too attractive for children and young people, and even for many of the "digital immigrants" charged with their upbringing and education (in the broadest sense). Adults and adult institutions have in effect lost much of the authority and will to force or induce children, adolescents, and college students to persevere with learning activities that are not immediately rewarding or stimulus-driven. If parents, educators, children, adolescents, and young adults do make an effort to resist the pull of "virtual reality," they are pitted against powerful commercial interests and a technological crusade aimed at turning all "users" into digital junkies.[39] As I already noted, these forces are now invading schools, education more generally, and life at all levels, without facing broad resistance. Moreover, they are frequently welcomed as the harbingers of forms of learning and student engagement best suited to the 21st century (despite the lingering concerns of many teachers).

As I see it, a public policy response to the current crisis in learning, reading, and neurosomatic development would require that the capitalist and technological juggernaut which feeds on existential burnout and a meta-addiction to overstimulation be defanged and tied down. But it can hardly be, barring some larger technological, social or environmental catastrophe—at which point the time for effective collective action may already have passed. Hopefully, deep reading, engaged learning, interest in larger social issues, and more sophisticated forms of thinking will not be extinguished as predicted by *Idiocracy*, the provocative dystopian movie released a decade ago. But these are likely to become niche preoccupations practiced by a relatively small reading elite mentally detached from the social mainstream.

As Birkerts warned over two decades ago, "the overall situation is bleak and getting bleaker."[40] He made this grim observation with reference to the kind of reading he cherishes, but it can also be applied to education in the broadest sense. With large-scale social institutions around the world facing a general crisis, it is hard to imagine how education can become the one effective island in a largely dysfunctional sea. As German sociologist Ulrich Beck has noted, we now live in an age when individuals are impelled to seek "biographical solutions" to deep systemic problems.[41] Or when, as a mock Soviet slogan proclaimed back in the 1920s, "the rescue of the drowning lies in the hands of the drowning themselves."[42]

What then can "we" (as parents, caregivers, and educators), as well as young people, do in order to stay sane and push back against the influence of the overwhelming social and technological forces I have described? The first step would be to grasp the predicament we all, and our children in particular, face. Unfortunately, the neurosomatic adaptations our sociotechnological milieu demands make this an almost impossible, and largely unwanted, task. What seems true, real or significant depends much on the degree to which emotional and visceral attunement infuses higher-order reasoning. And it is this existential involvement with the larger social world and "significant others" that seems most severely weakened as we speak, giving way to a technologically induced altered state of consciousness.

What would I advise readers who can still take seriously my broodings, or sense in them a (perhaps stronger) echo of their own doubts and struggles? I would first call on them to show all the determination they can muster and, if they have not already made some steps in this direction, stop spending most of their waking hours immersed in digital imagery and information—so they can break out of the vicious circle of virtualization of their own existence. This would also help them make a crucial effort to invest their relationships with the children and young people in their care with sufficient warmth. The rapid rise of "social media" has made maintaining this essential emotional bond much harder but even more imperative—a point made eloquently by Dr. Gabor Maté and developmental psychologist Gordon Neufeld.[43]

As I know from personal experience, this is easier said than done. I am someone who does get an easy high from reading an impressive feature article or non-fiction book. I am also keenly aware of, and spend much time ruminating on, the dangers I have described. Yet I have struggled to achieve in my own life a semblance of the balance I am advocating. Fortunately, my wife and I are blessed with a daughter who is extremely impressionable and eager to love us back even when we fail to give her the attention he seeks. She is also easily excited by seemingly trivial aspects of the "real world"[44]; is strongly attached to her grandparents; seeks out a limited number of deep friendships; and has been an exceptional student at a demanding high school (praised not just for her academic achievements, but also for her radiant presence in the classroom and willingness to help others). Gali is not glued to her phone texting, has not begged for an internet connection on it or for a tablet, and in the summer of 2014 spent willingly over a month (in several installments) without any internet access. Yet she also finds it difficult to follow consistently my unrelenting admonitions to limit the time she spends in front of her laptop (mostly searching information on various issues related—or often unrelated—to school assignments and planning her college education).

In my own teaching, I try to structure courses like *The Matrix*, providing a mixture of intellectual and more immediate stimulation. I use a mix of articles and book chapters—some academic, some potentially more engaging (for example, Michael Lewis's reports on the financial crisis—one of which provided material for *The Big Short* movie[45]), some in-between (like essays from *Foreign Affairs, Foreign Policy, The Atlantic,* and other high-brow publications). I usually select texts which address some larger issues related to structural or cultural trends that are often concealed behind the coverage of current events and developments—for example, changing understandings of freedom and power in the writings of major thinkers, or various implications of the rise of the market economy since the late 18th century. I also introduce some narrower concepts (for example, different explanations for extreme nationalism and related atrocities), and use bare-bone PowerPoint outlines to provide a degree of structure to classes and the whole course.

While I dearly hope such intellectual exploration can be sufficiently stimulating for some of my more curious students, I also include in my class sessions a few "action" sequences. I still show some brief video segments or distribute short articles and other handouts we can read and discuss in class. These are intended to provide more vivid illustrations or interpretations of the larger issues addressed in assigned readings, and to offer an opportunity for some more engaging activities—particularly for students who would otherwise be bored. I also ask students to write frequently, from one-paragraph responses they compose individually or in small groups in class, to longer integrative papers. Luckily, few students use laptops in the classroom, so I just banish them to the back (so they would distract others less), as opposed to any outright ban.

While I do make an effort to expose students to this learning mix, I still believe that the road to more sophisticated thinking and better writing passes mostly through "deep reading" and the neurophysiological maturation it spurs. In addition to selecting sufficiently complex yet emotionally engaging texts meant to facilitate this process, I also try to alert students to the indispensability of "real" reading, and the need to limit the time they spend online. More generally, I prod them to recognize the challenges in dealing with the information overload we all face. My overall goal is to allow and stimulate students to tune to gradually increasing levels of intellectual and moral complexity. I also try to make evident the benefits of further intellectual growth by exposing them to more sophisticated forms of thinking and making connections, sometimes between seemingly unrelated issues—as demonstrated in readings and comments from more intellectually advanced students (and sometimes from me).

Developing such an ability to think in context, relate concrete facts and developments to broader concepts and trends, and connect at a personal level to larger social issues is a sine qua non for achieving the goals of liberal education. Without such an overall aptitude, the knowledge students acquire can hardly become cumulative and "transfer" to new problems. Nor can they develop a more holistic understanding of the world and a capacity for lifelong, self-directed learning. Since I believe these abilities are primarily a matter of neural maturation and the more intricate forms of intellectual and emotional processing associated with it, I try to introduce students to some basics of brain development. I also give them tips for brain and physical fitness. I try to discuss these problems and offer illustrations mostly in good humor, as I am weary of offending students by implying that they are, as Mark Bauerlein put it in the provocative title of his book, just "dumb."[46] I try to frame issues related to brain development as "areas for improvement," and to emphasize the potential benefits of the mental gymnastics students can do in and outside of class. I also try to reach out to individual students who show curiosity, intellectual humility, and potential as earnest readers.

Despite all my efforts, too many of my students remain skeptical and aloof as I try to take them on this mental tour. Ultimately, the success of all as lifelong learners will depend on their own willingness and ability to muster sufficient intellectual and emotional resources to move on on their own. I do wish them much luck, since in the dizzying world we inhabit this has become an almost Herculean task. As I noted at the outset, it appears that many of my students are facing long odds—and they hardly seem exceptional in this respect. Personally, I am very much opposed to efforts to "disrupt" and "unbundle" the liberal arts model of higher education. Yet, it may well turn out that the majority of college students cannot fully benefit from this model and develop the holistic understanding and critical stance it is intended to foster.

My deeper worry is that we all—as individuals, parents, teachers, scholars, students, workers, entrepreneurs, leaders, etc.—have reached a point where we have become too jaded, even delusional as a result of the information and existential overload we suffer.[47] So we cannot collectively recognize—with sufficient clarity and urgency—the neurophysiological crisis we are facing; to say nothing of mustering the almost inhuman determination and human commitment needed to deal with it.

This is the recurring nightmare I have been unable to get out of my mind. It evokes a form of "technocracy" which, in the words of Scottish information society theorist Alistair Duff, should not be interpreted merely as "the rule of experts." Rather, it is a regime that involves "the rule of information technology, the domination of information technology over human beings, and the subordination of people to a technological imperative."[48] The kind of "technocracy" Duff describes seems to shape too many of us into restless cogs whose labor and life energy can be almost constantly tapped, and who are engrossed in abundant forms of self-stimulation. We are thus assisting the technocratic system in which we are caught to expand its tickling tentacles into every crevice of our social and personal lives—while often dreaming dreams of freedom and self-empowerment. This is a role for which "the best and the brightest," or the cognitive 1%, are richly rewarded—until one day even their input is perhaps rendered obsolete.[49]

Over eight decades ago, Aldous Huxley dreamed up a most frightening future society. In it, human fetuses are manipulated prior to their artificial "hatching" in order to stunt the natural development of some and thus create castes of individuals fitted for different social roles.[50] They live lives devoid of strong attachments or emotions, saturated with opportunities to instantly gratify various desires—so these will not be pent up to the point of bursting and causing social disruption. Suspended in this milieu, most individuals feel free and happy—unlike previous generations or the recalcitrant "savages" on the edge of "civilization."

In my nightmare scenario, the sociotechnological "matrix" we inhabit has developed a more subtle way of shaping our proclivities, abilities, and attitudes. It achieves this by providing sources of chronic stimulation, tapping the exceeding plasticity of the human brain, and provoking longer-term neurophysiological modification. As English professor Edward Mendelson notes, "Virginia Woolf's serious joke that 'on or about December 1910 human character changed' was a hundred years premature. Human character changed on or about December 2010, when everyone, it seemed, started carrying a smartphone."[51] This was perhaps the tipping point in a longer-term process of adaptation to social and sensory overstimulation. As a result of this adjustment, we have progressively turned into "mental penguins"—beings who have shed some deeply human qualities in order to acquire a more adaptive yet narrower "skill set."

Huxley Was Right?

Over three decades ago, Neil Postman was nagged by a troubling thought—"the possibility that Huxley, not Orwell, was right"[52] in describing the totalitarianism of the future. He introduced his book, *Amusing Ourselves to Death*, with a premonition that ran counter to the spirit of Ronald Reagan's "morning in America":

> What Orwell feared were those who would ban books. What Huxley feared was that there would be no reason to ban a book for there would be no one who wanted to read one. Orwell feared those who would deprive us of information. Huxley feared those who would give us so much that we would be reduced to passivity and

egoism. Orwell feared that the truth would be concealed from us. Huxley feared the truth would be drowned in a sea of irrelevance. Orwell feared we would become a captive culture. Huxley feared we would become a trivial culture, preoccupied with some equivalent of the feelies, the orgy porgy, and the centrifugal bumblepuppy.

As Huxley remarked in *Brave New World Revisited,* the civil libertarians and rationalists who are ever on the alert to oppose tyranny "failed to take into account man's almost infinite appetite for distractions." In *1984,* Huxley added, people are controlled by inflicting pain. In *Brave New World,* they are controlled by inflicting pleasure. In short, Orwell feared that what we hate will ruin us. Huxley feared that what we love will ruin us.[53]

If only Huxley had realized that what we crave most may be information as such.[54] In any case, since the time of his and Postman's dire warnings, the information-carrying caravan has only picked up speed.

I am under no illusion that what I write can make the smallest dent in the larger logic of the zeitgeist. The musings of skeptics like Neil Postman, Jane Healy, Sven Birkerts, Nicholas Carr, or Susan Greenfield, or more balanced coverage of our digitally-enhanced mental environment (for example, *The New York Times* series "Your Brain on Computers" published in 2010) have obviously failed to achieve this. So I can hardly hope to do better, or even to be part of some larger awareness-raising effort.

As Gabor Maté has noted, childhood development problems and compulsive behaviors are hardly a matter of personal or parental failure. Rather, they reflect "a social and cultural breakdown of cataclysmic proportions."[55] Alas, I cannot imagine a set of institutional or technological adjustments that can reverse this seismic shift. The only realistic form of resistance I envision is an effort to create small sanctuaries for ourselves and our loved ones; and reach out within and out of these to a few kindred souls. It is particularly important to reach out, before it is too late, to the young learners in our care (in the broadest sense of these words)—and help them grow as best as we can. So they can develop a fuller grasp of the "matrix" they inhabit—and of their own unique mix of strengths and vulnerabilities.

My more upbeat critics will no doubt dismiss all these premonitions as the flawed arguments of an uninformed amateur or the alarmist delusions of a troubled mind—or both. They will point out that worries about technological change have always been proven wrong—as humankind has adapted to technological change by developing new and leaving behind some outdated aptitudes. It may appear that even real penguins in Antarctica do not quite miss the stereotypical bird traits they have lost. Until perhaps a giant iceberg cuts them off from the ocean,[56] or global warming threatens their food supply.[57] For our children's sake, I hope against hope that the optimists are at least partly right.[58]

Notes

1 My (rather literal) translation from the Russian original.

1. The End of Education Revisited

1 See Arendt, "The Crisis in Education."

2 My excessive use of quotation marks reflects my inability to take at face value many common terms.

3 Postman and Weingartner, *Teaching as a Subversive Activity.*

4 Postman, *Teaching as a Conserving Activity.*

5 Postman, *Amusing Ourselves to Death: Discourse in the Age of Show Business.*

6 Postman, *The End of Education: Redefining the Value of School.*

7 My excessive use of passive voice reflects a less egocentric mode of thinking and expression which is common to Slavic and many other languages.

8 See, for example, Nussbaum, *Not for Profit: Why Democracy Needs the Humanities*; Delbanco, *College: What It Was, Is, and Should Be*; Small, *The Value of the Humanities.*

9 For an extreme example, see Rhee, *Radical: Fighting to Put Students First.*

10 See Posner and Rothbart, *Educating the Human Brain*; Wolfe, *Brain Matters: Translating Research into Classroom Practice*; Sousa et al., *Mind, Brain, and Education: Neuroscience Implications for the Classroom.*

11 See Lehrer, "The Truth Wears Off." Unfortunately, Lehrer was disgraced when a vigilant journalist revealed that he had invented Bob Dylan quotes for his latest book, *Imagine: How Creativity Works.* He was also blamed for recycling some of his writing, sloppy research, and misrepresenting findings in neuroscience and medical research (see Kachka, "Proust Wasn't a Neuroscientist. Neither Was Jonah Lehrer."). As a result, publishers pulled *Imagine* and a

previous book from bookstores. Though Lehrer's lapses of judgment and integrity are inexcusable, I still believe some of his insights into the broader implications of neuroscientific research (and on the limitations of the scientific method in general) can be illuminating.

12 See Tallis, *Aping Mankind: Neuromania, Darwinists and the Misrepresentation of Humanity*.

13 See "The Rich, the Poor and Bulgaria."

14 Quoted in McDonald, "Coming to America."

15 Tocqueville, *Democracy in America*, vol. 1, ch. xv, "Unlimited Power of the Majority in the United States, and Its Consequences."

16 See Baudrillard, *America*; Eagleton, *Across the Pond: An Englishman's View of America*.

17 Except for the frequent passive constructions which I cannot quite resist. The same applies to the "buts" and "ands" at the start of many sentences, and some other linguistic oddities. Though I understand these are considered stylistically substandard in English, they partly reflect my own more convoluted thinking process. They are also quite typical of the Slavic languages I am familiar with—which are generally less egocentric and enmeshed with a more fatalistic worldview.

18 A more apt comparison here would be with Evgeny Morozov, the fierce critic of IT evangelism and algorithmic solutions to complex social problems (see *To Save Everything, Click Here: The Folly of Technological Solutionism*). He was a student of mine at the American University in Bulgaria, but I can claim no credit for his evolution as an intellectual and a skeptic regarding the alleged liberating potential of information technology. We had not been in touch since he graduated in 2005. When we met again in the spring of 2012, he struck me as having stronger faith in the empirical verification of even the broadest theories and in evidence-based

policy solutions.

19 See Sussman, "Mental Illness and Creativity: A Neurological View of the 'Tortured Artist.'" When I read this article a few years ago, it immediately struck me that the fine-tuned integration of emotional and logical processing in my own brain may have been knocked off balance by a few mild concussions I suffered as a child. The prefrontal cortex, which plays a vital role in orchestrating the potential cacophony of neural signaling in the brain (see Goldberg, *The Executive Brain: Frontal Lobes and the Civilized Mind; The New Executive Brain: Frontal Lobes in a Complex World*) and the neural fibers which connect it to other brain areas, are particularly vulnerable to even mild traumatic shocks. And even minor disruptions in the way the prefrontal cortex coordinates the brain's neural ensemble can produce marked changes in personality, emotional attunement, and outlook. These problems have recently been brought into the public spotlight by experts and journalists discussing the dangers of concussions in American football, European football or soccer, boxing, extreme sports, etc. (particularly for children and teenagers).

20 I owe a particularly large debt to the numerous neuroscientists who have had the ingenuity and tenacity to design and see through thousands of clever experiments with lab animals and human subjects. I could have never conducted such research myself, even if I had the training—particularly experiments involving any surgery on living organisms. Though such experiments have sometimes produced truly insightful results, it is clearly not for the faint at heart.

21 Westen, *The Political Brain: The Role of Emotion in Deciding the Fate of the Nation*; Haidt, *The Righteous Mind: Why Good People Are Divided by Politics and Religion.*

22 This is how Sven Birkerts, one of my contemporary cultural heroes, describes "the central paradox facing the writer who

would still try to address the present-day world in significant terms": "the more faithful you are to your truth, the more deeply you see into the dynamics of what is taking place on all sides, the less of a chance there is that your version of things will get published, or, if published, bought and read" (*The Gutenberg Elegies: The Fate of Reading in an Electronic Age*, 207). Though Birkerts makes this observation with reference to a few modern writers, it could be applied to gloomy non-fiction, too (though his books—and other potentially disturbing prophesies—have been published by major imprints and have been much acclaimed). Applying the same maxim to my own musings borders on grotesque self-flattery—still I can hardly resist it.

23 There are easy ways to determine with sufficient precision if your own mental wavelength is on a different band from mine, if this is not already obvious. If Nicholas Carr's provocative question "Is Google Making Us Stupid?" or Susan Greenfield's warnings about the perils of a "screen-based lifestyle" (*Mind Change: How Digital Technologies Are Leaving Their Mark on Our Brains*, 19) do not resonate with some of your concerns, probably nothing I say further will. I will try to explain later where this divergence of sensibilities and existential anxiety—or lack thereof—may come from.

24 A few of these boxes are based on entries in a blog I have maintained over the last couple of years (sardamov.blogspot.com).

2. College Teaching and Its Discontents

1 Until I graduated in 1998, the department I loved so much bore an old-fashioned name—Department of Government and International Studies. That name was quite appropriate since it reflected the more traditionalist approach to scholarship I found there. Later, a team of external evaluators recommended a more conventional name—Department of

Political Science. The department took this advice, and has since moved on to hire more faculty with an empiricist bent as it has tried to compete against research powerhouses and rise up in academic rankings.

2 See Kaplan, "The Coming Anarchy"; "Was Democracy Just a Moment?"

3 See Goldhagen, *Hitler's Willing Executioners: Ordinary Germans and the Holocaust*.

4 Agha and Malley, "The Last Negotiation: How to End the Middle East Peace Process."

5 In recent two-hour exams, a few graduating student have written three-four-page essays (A4 format) comprised of one such "paragraph."

6 Dalrymple, "The Case for Cannibalism."

7 In January 2012, as we were discussing an assigned reading on Pakistan's tribal areas in my "Conflict and Conflict Resolution" class, I asked students if they knew what form of government the country had. No one had a clue, and several guessed it was some kind of dictatorship. At that point, no one had read or heard of "drones" either.

8 This overconfidence could reflect the now famous Kruger-Dunning effect—a cognitive bias which makes it difficult for incompetent individuals to recognize their own lack of knowledge or ability (see "Why People Have So Much Trouble Recognizing Their Own Incompetence"). Such a tendency among Bulgarian students is a bit surprising since Bulgarians tend to habitually underestimate their own abilities, achievements, and social status (see Sgourev, "Lake Wobegon Upside Down: The Paradox of Status-Devaluation").

9 See Edmundson, "The Uses of Liberal Education: 1. As Lite Entertainment for Bored College Students."

10 English professor Ranjan Adiga offers a particularly unsettling example of what he sees as an "empathy vacuum" in

American college classrooms. The morning after a destructive earthquake struck Nepal, his native country, he was disheartened by the reaction—or lack thereof—among his students at a liberal arts college in Salt Lake City. They had not followed the news, and showed little interest or sympathy when he mentioned the terrible disaster (see Adiga, "Even an Earthquake Can't Stir Student Empathy").

3. Neuroscience Rides to the Rescue

1 *Brain Story* was the title of a popular documentary series that aired on BBC in 2000. It was narrated by renowned, if sometimes controversial, British neuroscientist Baroness Susan Greenfield.

2 See Herbert, "Sophistry or Sensitive Science?" Herbert later presented her ideas more comprehensively in a book she co-wrote with Karen Weintraub, *The Autism Revolution: Whole-Body Strategies for Making Life All It Can Be.*

3 See Jill Neimark's interview with immunologist and environmental health expert Claudia Miller, "Is The World Making You Sick?"; Bennett et al., "Project TENDR: Targeting Environmental Neuro-Developmental Risks. The TENDR Consensus Statement."

4 See Ahlbom and Feychting, "Electromagnetic Radiation: Environmental Pollution and Health."

5 Pearce, *Evolution's End: Claiming the Potential of Our Experience.*

6 Zull, *The Art of Changing the Brain: Enriching Teaching by Exploring the Biology of Learning.*

7 Shenk, *Data Smog: Surviving the Information Glut.* The growing frequency of episodes similar to the ones I described above made Shenk's book resonate forcefully.

8 The term "information overload" was coined by futurologist Alvin Toffler in the early 1970s. See Toffler, *Future Shock.*

9 Sardamov, "Teaching and Learning in the Age of

Audiovisual Pollution."

10 American University in Bulgaria, "Fulfilling the Promise: A Strategic Plan for the American University in Bulgaria, 2010–2015."

11 See Brown, Roediger III, and McDaniel, *Make It Stick: The Science of Successful Learning*.

4. The Limits of "Neuroeducation"

1 This is a graduation exam mandated under Bulgarian law, a less daunting version of similar exams in the Soviet and particularly the German education systems. It is taken by Bulgarian students and many others in whose native countries a Bulgarian diploma is more easily recognized by the education authorities than an American one (former Soviet republics adhere to the agreement for the mutual recognition of academic degrees Bulgaria once signed with the Soviet Union).

2 Freeland, "The Rise of the New Global Elite."

3 Bremmer and Roubini, "A G-Zero World: The New Economic Club Will Produce Conflict, Not Cooperation."

4 My term.

5 This confusion has persisted as I have subsequently asked students to discuss the same question and produce a written response in small groups in class.

6 See Bauerlein, *The Dumbest Generation: How the Digital Age Stupefies Young Americans and Jeopardizes Our Future (Or, Don't Trust Anyone Under 30)*.

7 Cf. Edmundson, "Uses of Liberal Education."

8 Following the accession of Bulgaria and Romania to the European Union in 2007, more Bulgarian students are choosing to study at West European universities, and AUBG has lost all attraction for Romanian students. Higher numbers of young people from other Southeast European countries are also looking West as many newly-hatched

English-language programs (which often charge zero or nominal tuition) are clamoring to attract qualified applicants. As a result, AUBG has needed to become a bit less selective.

9 Lehrer, "The Truth Wears Off."

10 Edmundson, "Uses of Liberal Education."

11 Bauerlein, *Dumbest Generation*. I am wondering how many of the American students in Bauerlein's English classes would immediately recognize the historical associations in his title, and in the subtitle which includes the phrase "Don't Trust Anyone Under 30." In an upper-level class on "Culture and Power" I taught in the spring of 2014, no one out of 26 students had come across the phrase Bauerlein mocks—though one guessed right when I hinted that it was a famous slogan from the 1960s youth rebellion.

12 See Shenkman, *Just How Stupid Are We? Facing the Truth About the American Voter*.

5. The End of Authority

1 It is quite common to see groups of students smoking fairly close to the main entrances of high schools, and alcohol consumption sometimes takes place after classes just a bit farther away. In June 2011, we took Gali to visit one of the most prestigious high schools in Sofia which we were considering for her. The grounds just outside the school's fence were strewn with cigarette buts, empty cigarette packs, and beer bottles.

2 See "PISA: Bulgarian Students Come Last in Europe"; Resmovits, "And the World's Worst Problem Solvers Are..."

3 See Thompson, *Moral Panics*.

4 See Lazzeri, "Dad's Not Alfie."

5 See, for example, Barrett, "Anti-Social Behavior Growing, Says Official Survey"; Seligson, "Cutting: The Self-Injury Puzzle"; Barrionuevo, "In Tangle of Young Lips, a Sex

Rebellion in Chile"; Sternbergh, "Up with Grups."

6 Rosen, "The Overpraised American."

7 Veno and van den Eynde, "Moral Panic Neutralization Project: A Media-Based Intervention."

8 Barron and Lacombe, "Moral Panic and the Nasty Girl."

9 Ibid., 64.

10 Ibid., 65.

11 See Welch, "Heroin Epidemic? Not So Fast, Says Carl Hart"; "Drug Legislation: On Bongs and Bureaucrats."

12 Heffernan, "When the Cravings Won't Quit, Turn On the Camera."

13 See Westen, *Political Brain*; Haidt, *Righteous Mind*; Hibbing, Smith, and Alford, *Predisposed: Liberals, Conservatives, and the Biology of Political Differences*. I use the term "neurosomatic" to point to the unity of brain and body. While the workings of the human mind are not reducible to neural processing, I do believe mental "operations" depend on how the human brain functions, and how it is integrated with the body. More on this in chapter 10.

6. My Stroke of Insight

1 That evening, I might have had an extra drink or two; or perhaps I was tired after a day of teaching and commuting. There is some research indicating that these conditions could unchain creativity. According to the theory explaining this correlation, when the external focus and inhibitory powers of the prefrontal cortex are weakened, the brain becomes capable of forming more far-flung associations (see Smith, "Creativity and IQ, Part I: What Is Divergent Thinking? How Is It Helped by Sleep, Humor, and Alcohol?"). Weaker cognitive control and self-discipline could thus help you become more creative—and in some cases die from an overdose of drugs or alcohol, or succumb to serious mental illness. Sleepiness (particularly when half-

awake early in the morning) can also disrupt prefrontal activity—providing a safer creativity boost, as long as sleep-deprivation does not become chronic.

2 See Malamed, "Emotion and Learning."

3 See Daniels, "Living with Intensity: Overexcitabilities in Profoundly Gifted Children." David Dobbs has popularized a related theory, the "orchid hypothesis" (see Dobbs, "The Science of Success").

4 Friedman, "Lasting Pleasures, Robbed by Drug Abuse."

5 See Kang, "The High Is Always the Pain and the Pain Is Always the High."

6 See Carr, "Is Google Making Us Stupid?"; *The Shallows: What the Internet Is Doing to Our Brains*.

7. A Well-Tempered Brain

1 Neuropsychiatrist Peter Whybrow refers to a "well-tuned brain" in the title of his latest book (see *The Well-Tuned Brain: Neuroscience and the Life Well Lived*. He explains he chose this phrase with reference to Johan Sebastian Bach's famous collection, "A Well-Tempered Clavier." I had settled on "well-tempered brain" before I read Whybrow's book, and decided to keep it as I find it a bit more resonant.

2 By all accounts, the behavior I found so shocking at the conference has fast become common. For example, during the proceedings of the Breivik trial in Oslo, one of the five judges was caught by a camera playing solitaire on his laptop (see "Breivik Judge Caught Playing Solitaire in Court"). This kind of multi-tasking seemed particularly striking because of the nature of Breivik's crime and the harrowing testimony offered during his trial. A spokes-woman for the court explained that the judges were, indeed, attentively following the proceedings. "There are different ways to stay focused," she added.

3 Carr, "Keep Your Thumbs Still While I'm Talking to You."

The participants, mostly young men, were IT enthusiasts and early adopters. But in recent years participants in academic conferences in the humanities and social sciences have similarly been encouraged to live-blog, tweet, tune into streamed sessions, etc.

4 Stone, "Continuous Partial Attention."

5 A few years ago I read about an IT consultant who bragged how he always carried with him several electronic devices dedicated to different purposes. He said he was in almost constant communication with co-workers, family, and friends; and he regularly updated several blogs geared towards those different professional and social circles. That level of digital interaction struck me as weird at the time. But now someone like this would also be tweeting and calling up different apps throughout the day. And, as the saying goes, this is already close to becoming the new normal.

6 See Young, "Internet Addiction: The Emergence of a New Clinical Disorder"; Aboujaoude, *Virtually You: The Dangerous Powers of the E-Personality*; Christakis, "Internet Addiction: A 21st Century Epidemic?"

7 See Seib and Dayton, "When Games Start Playing You: Cyber Addicts"; Siew, "U.S. Students Suffering from Internet Addiction." The notion of "internet addiction" may have initially cropped up as a satirical hoax (see Beato, "Internet Addiction"), but has recently received more traction. Of course, there are many doubters (see Civita, "Internet Addiction as New Addiction?"; Frances, "Internet Addiction: The Next New Fad Diagnosis"; Heffernan, "Case of Internet Addiction").

8 As already indicated, the ur-text describing these concerns is Carr, "Is Google Making Us Stupid?"; see also Birkerts, *Gutenberg Elegies*; Birkerts, "Reading in a Digital Age"; Ulin, *The Lost Art of Reading: Why Books Matter in a Distracted Time.*

9 See Seib and Dayton, "When Games Start Playing You."

10 See "S. Korea Child 'Starves as Parents Raise Virtual Baby.'"

11 The 2013 documentary, *Web Junkie*, describes the problem of internet addiction in China and the ordeals of adolescents boys sent to a rehabilitation center in Beijing.

12 See Young, *Caught in the Net: How to Recognize the Signs of Internet Addiction—and a Winning Strategy for Recovery.*

13 See Dunckley, *Reset Your Child's Brain: A Four-Week Plan to End Meltdowns, Raise Grades, and Boost Social Skills by Reversing the Effects of Electronic Screen-Time.*

14 According to a recent study, just a few days of digital abstinence in a natural setting can noticeably improve the ability of sixth graders to read emotions (see Uhls et al. "Five Days at Outdoor Education Camp Without Screens Improves Preteen Skills with Nonverbal Emotion Cues"). More radical approaches to "resetting" one's dopamine system may be even more effective, but would require awareness of the deficits induced by chronic overstimulation and much willpower—precisely the mental resources that are depleted the most by compulsive self-stimulation (see Moyer, "Overstimulation and Desensitization—How Civilization Affects Your Brain"; Becker, "Change Your Receptors, Change Your Set Point").

15 See Newman and Harris, "The Scientific Contributions of Paul D. MacLean (1913–2007)." Though MacLean's theory of the "triune brain" has recently been criticized as overly simplistic, it may provide a crucial insight into the basic, fairly inelegant, design of the human brain. More detailed models, on the other hand, may have their own problems. More on that later.

16 This term was introduced by MacLean. It has also been criticized as simplistic, since the brain centers he described as forming the "limbic system" do not seem to form a distinct functional ensemble; and it has become obvious that parts of the neocortex play a key role in the processing of emotions

(see LeDoux, "Parallel Memories: Putting Emotions Back into the Brain"). Yet, identifying this separate "limbic system" as part of the triune brain may serve a larger purpose, despite any perceived lack of neuroscientific precision.

17 For example, the cerebellum plays a key role in sustaining attention, and some neuroscientists believe the "reptilian brain" may act as the brain's "spam filter" by helping block out irrelevant stimuli (see McCollough and Vogel, "Your Inner Spam Filter"). As it will become clear, the release of dopamine in the midbrain has a profound effect on the prefrontal cortex and other parts of the brain.

18 See Goldberg, *Executive Brain*; *New Executive Brain*.

19 Quoted in Blakeslee, "In a Host of Ailments, Seeing a Brain Out of Rhythm."

20 Whybrow, *American Mania: When More Is Not Enough*, 62–3.

21 See McNamara, "The God Effect."

22 See Yoffe, "Seeking: How the Brain Hard-Wires Us to Love Google, Twitter, and Texting. And Why That's Dangerous."

23 See Fleming, "The Science of Craving."

24 Other researchers have similarly come to the conclusion that an excess of dopamine serves to produce not "pleasure," but heightened arousal and motivation for exploration and learning. This tendency is illustrated by the typical profile and behavior of individuals diagnosed with schizophrenia, a condition marked by chronically elevated dopamine levels in the brain. As psychologist Richard Katz once commented, if the "pleasure" hypothesis was valid, such patients should be "inextinguishably fat and happy"; instead, "they are just the opposite" (quoted in Wickelgren, "Getting the Brain's Attention").

25 Lehrer, "The Itch of Curiosity"; cf. Biederman and Vessel, "Perceptual Pleasure and the Brain."

26 See Blakeslee, "A Small Part of the Brain, and Its Profound

Effects."

27 See Damasio, *Descartes' Error: Emotion, Reason, and the Human Brain*.

28 See Wickelgren, "Getting the Brain's Attention."

29 See Pert, *Molecules of Emotion: Why You Feel the Way You Feel.*

30 See Pearce, *Evolution's End*.

31 See Gershon, *The Second Brain: The Scientific Basis of Gut Instinct and a Groundbreaking New Understanding of Nervous Disorders of the Stomach and Intestine*.

32 See Peretti, "Why Our Food Is Making Us Fat."

33 See Almedrala, "The Surprising Things Exercise Can Do for Your Brain."

34 Recent research has demonstrated that sleep is essential for the formation of long-term memories. According to one theory, during deep, slow-wave sleep the hippocampus reactivates emotionally tagged memories from the preceding waking hours. Replaying such memories facilitates their long-term storage in the neocortex (see Begley, "How to Improve Your Memory with Sleep"). According to recent research, just one sleepless night can affect the expression of genes controlling the human body clock (see Dockrill, "Just One Night of Sleep Loss Can Alter Our Genes, Study Finds").

35 See Lubin et al., "Epigenetic Mechanisms: Critical Contributors to Long-Term Memory Formation."

36 See Hood, "Re-creating the Real World."

8. A Fatal Attraction

1 For a real-life example of someone who has "monetized" even more spectacularly his brilliance as a mathematician, see Broad, "Seeker, Doer, Giver, Ponderer: A Billionaire Mathematician's Life of Ferocious Curiosity."

2 See Rosen, "The Age of Egocasting."

3 See Wickelgren, "Getting the Brain's Attention."

4 See Robbins, "Missing the Big Picture."

5 See Bechara et al., "Decision Making, Impulse Control and Loss of Willpower to Resist Drugs: A Neurocognitive Perspective."

6 Ibid.

7 Ibid.; cf. Friedman, "Pleasures Robbed by Drug Abuse." Adolescents are particularly susceptible to such ill-advised pursuits since they tend to experience excitements more powerfully, while the reflective system in their brains is still underdeveloped (see Dobbs, "Beautiful Brains").

8 See Rasheed and Alghasham, "Central Dopaminergic System and Its Implications in Stress-Mediated Neurological Disorders and Gastric Ulcers: Short Review."

9 See Bechara et al., "Loss of Willpower to Resist Drugs"; Friedman, "Pleasures Robbed by Drug Abuse." The warped perspective of heavy drug addicts can sometimes lead them to commit grotesque crimes. In the past, such crimes were associated mostly with illegal drugs. More recently, the abuse of illegal substances has been supplemented by addiction to prescription drugs. This shift has resulted in a spike of pharmacy robberies. In some states, these have even outpaced the more traditional bank robberies. While some of the hold-ups are carried out by well-organized gangs, others are done by desperate addicts (see Goodnough, "Pharmacies Besieged by Addicted Thieves").

10 See Fields, "Money Buys Unhappiness, Proven in a New Study"; Seltzer, "Greed: The Ultimate Addiction."

11 See Friedman, "Pleasures Robbed by Drug Abuse."

12 Ibid.

13 Kang, "The High Is Always the Pain."

14 As the Wikipedia entry states (as of July 6, 2016), "behavioral addiction has been proposed as a new class in DSM-5, but the only category included is gambling addiction. Internet gaming addiction is included in the appendix as a condition for further study."

15 With the morphing of many "investment" techniques into high-stakes gambling, a similar desensitization and loss of perspective can perhaps help explain the reckless pursuit of ever higher financial pay-offs that led to the current financial debacle. See Friedman, "A Crisis of Confidence for Masters of the Universe."

16 See Cash et al., "Internet Addiction: A Brief Summary of Research and Practice."

17 See Bechara et al., "Loss of Willpower to Resist Drugs"; Sample, "The New Pleasure Seekers."

18 In 2007, John Paulson's firm managing several hedge funds made 15 billion dollars—a sum which seems to defy the human imagination. Interestingly, he never entertained the obvious thought that maybe that kind of reward for his foresight was enough, and he could retire and spend the rest of his life in leisure and luxury.

19 See Yoffe, "Seeking."

20 See Knight, "'Info-Mania' Dents IQ More Than Marijuana."

21 See Young, "Internet Addiction"; Aboujaoude, *Virtually You.*

22 See Anderson, "Just One More Game."

23 See Greenfield, *i.d.: The Quest for Identity in the 21st Century,* 199–202.

24 See Zuder, "A General in the Drug War"; Richard A. Friedman, "Who Falls to Addiction and Who Is Unscathed?"; cf. Pulvirenti and Diana, "Drug Dependence as a Disorder of Brain Plasticity: Focus on Dopamine and Glutamate."

25 See Patoine, "Is 'Internet Addiction' a Psychiatric Disorder?"

26 See Heffernan, "Case of Internet Addiction."

27 Curiously, Heffernan has credited Twitter with ushering in a new golden age of ubiquitous poetry (see Heffernan, "Poetry in the Age of Poetry"). Her inability to cringe from almost anything is matched by journalist and writer Hanna Rosin. Chronically upbeat, she has argued (partly based on her own experience) against breastfeeding and imposing any limits

on the use of digital devices by children (see "The Case Against Breastfeeding"; "The Touch-Screen Generation"). She has also concluded that the "hook-up culture" is ultimately good for young women since it allows them to combine the pursuit of higher education and a successful career with sexual gratification until they decide to pursue (or forego) matrimony (see "Boys on the Side").

28 See Greenfield, *Mind Change*, 165–6, 170.

29 See Small and Vorgan, *iBrain: Surviving the Technological Alteration of the Modern Mind*.

30 For example, some studies have found that experienced gamers make better drone pilots and laproscopic surgeons (see Lasalle, "Gamers Make Awsome Drone 'Pilots,' Study Says (Be Afraid)"; Rosser et al., "The Impact of Video Games on Training Surgeons in the 21st Century").

31 See Mosher, "High Wired: Does Addictive Internet Use Rewire the Brain?"; Greenfield, *Mind Change*. There is some research indicating that similar changes in the brain can be induced even by physiological influences rarely associated with any form of addiction—for example, regular tanning (see O'Connor, "How Tanning Changes the Brain").

32 See Greenfield, *Mind Change*, 209.

33 See de Zengotita, "The Numbing of the American Mind: Culture as Anesthetic."

34 See Carr, "Is Google Making Us Stupid?"; *Shallows*; cf. Birkerts, "Reading in a Digital Age."

9. Dumb and Dumber

1 Arum and Roksa, *Academically Adrift: Limited Learning on Academic Campuses*.

2 The first thing students learn in an introductory social science methodology course is that correlation does not equal causation. In this particular case, it could well be that stronger, more intelligent and motivated students will

choose more rigorous courses and spend more time studying. In any case, many other studies had already confirmed that students were recently studying much less compared to earlier generations. One such survey found that between 1961 and 2003 the average time college students spent preparing for courses had declined from 24 to 14 hours per week (see Avila, "Children These Days Spend Less Time Studying and More Time Playing, Study Finds").

3 See Gelman, "Is 'Academically Adrift' Statistically Adrift?"; Sternberg, "Who Is Really Adrift?" Not taking a challenging test seriously, on the other hand, could be seen as symptomatic in its own right.

4 Professors who want to reduce the time and energy they commit to teaching can draw on ready-made tools provided by textbook publishers—banks of PowerPoint presentations, quiz and exam questions (including multiple-choice), assignments, etc. The growing numbers of adjuncts and graduate students teaching key courses may find it difficult to give each course and student adequate attention for wholly different reasons.

5 See Edmundson, "Who Are You, and What Are You Doing Here?"

6 Bauerlein, *Dumbest Generation*.

7 Ibid., 17–18.

8 See Cunningham and Stanovich. "What Reading Does to the Mind," 144–5.

9 See Intercollegiate Studies Institute, "The Coming Crisis in Citizenship."

10 McLuhan, *Understanding Media: The Extensions of Man*, 7. To dramatize this point, McLuhan paraphrased a quote by T. S. Eliot into an evocative metaphor. He compared the content of any mediated message to "the juicy piece of meat carried by the burglar to distract the watchdog of the mind" (32).

11 McLuhan and Zingrone, *Essential McLuhan*, 239.

12 See McLuhan, "The Brain and the Media: The 'Western' Hemisphere."

13 See Winn, *The Plug-In Drug: Television, Children, and the Family*.

14 Kubay and Csikszentmihalyi, "TV Addiction Is No Mere Metaphor." Csikszentmihalyi is the more famous of the two authors, best known for his theory of "flow"—a term indicating rapt absorption in an activity (see *Flow: The Psychology of Optimal Experience*).

15 Lewin, "If Your Kids Are Awake, They Are Probably Online."

16 Ibid.

17 See Richtel, "Wasting Time Is New Divide in Digital Era."

18 Coates, "Social Network Disconnect."

19 National Center for Educational Statistics, "The Nation's Report Card: Civics 2006 (National Assessment of Educational Progress at Grades 4, 8, and 12)."

20 National Center for Educational Statistics, "The Nation's Report Card: U.S. History 2006 (National Assessment of Educational Progress at Grades 4, 8, and 12)."

21 Gould, *Guardians of Democracy: The Civic Mission of Schools*, 18–19.

22 Torney-Purta, "Patterns of the Civic Knowledge, Engagement, and Attitudes of European Adolescents: The IEA Civic Education Study," 136.

23 See Bartunek, "Belgian Trainee Teachers Fail in Basic General Knowledge."

24 Romano, "Literacy of College Graduates Is in Decline."

25 Ibid.

26 Ibid.

27 Lauritsen, "Study: More College Students Need Remedial Classes."

28 Romano, "Will the Book Survive Generation Text?"; cf. Baron, "The Plague of tl; dr."

29 Crain, "Twilight of the Books: What Will Life Be Like If People Stop Reading?"

30 Ibid.

31 Ibid.

10. ...or a Neurosomatic Crisis?

1 See, for example, Willingham, "Is Technology Changing How Students Learn?"

2 See Giedd, "The Digital Revolution and Adolescent Brain Evolution."

3 The best known developmental psychologist who has argued along these lines is Harvard professor Jerome Kagan. See Kagan, *Three Seductive Ideas*.

4 Brain plasticity is most frequently understood as synaptic plasticity—the formation and long-term potentiation (or weakening and elimination) of synaptic connections between neurons. It is also related, however, to neurogenesis (the birth of new neurons) in key brain centers and the myelination of neural fibers or axons (their coating in white fatty sheathing which facilitates speedier and clearer neural transmission). Brain plasticity additionally includes other intricate changes in support cells, capillaries and blood flow, the synthesis of proteins, etc. All these modifications are influenced not only directly by experiences and neural signaling, but also by epigenetic changes. Moreover, the brain forms a functional whole with the body, and plasticity can thus involve adaptations at the level of the entire organism.

5 See Kagan, *In Over Our Heads: The Mental Demands of Modern Life*.

6 Khamsi, "Dolphins Play the Name Game, Too."

7 Griggs, "Border Collie Takes Record for Biggest Vocabulary."

8 In 2009, a male chimpanzee that had apparently been "socialized" at a family's house since birth (a process which

included appearances in TV shows and commercials) became involved in a notorious incident. Though he had been given antidepressants, he went berserk and severely mauled an approaching family acquaintance. His owner rushed out of the house and stabbed him repeatedly with a knife, but was unable to stop the attack. The raging animal was eventually shot and killed by a police officer responding to the 911 call (see Walsh, Bryan. "Why the Stamford Chimp Attacked").

9 See Kneissle, "Research into Changes in Brain Formation."

10 Cited in ibid.

11 A few decades later, individuals in possession of such a "new brain," or of even newer versions, form the bulk of German engineers. This could partly account for some embarrassing failings that have dulled a bit the prestige of German engineering—for example, the failure to launch the new Berlin airport into operation, the delayed delivery by Siemens of a major order of new trains for the German railways and Eurostar, and other problems (see Mangasarian, "Berlin Airport Fiasco Shows Chinks in German Engineering Armor"; Hawley, "Siemens Problems 'Can Hardly Get Worse'"; Holden, "German Cars 'Among Worst for Engine Failure'"). As the case of Volkswagen emissions cheating demonstrates, even competent engineering can cause tremendous problems when steered by managerial hubris. Such difficulties are unlikely to be limited to Germany, or to mechanical engineering and car sales—with too many decision-makers in politics, high finance, and "big data" companies demonstrating that the kids do not always end up alright.

12 Herbert, "Sophistry or Sensitive Science?"

13 Ibid.

14 In fact, physics teachers have long struggled to help high school and college students overcome this intuitive belief

(see Champagne and Klopfer, "Native Knowledge and Science Learning," 4).

15 See Allen, "A New International Business Language: Globish."

16 See Foroni, "Do We Embody Second Language? Evidence for 'Partial' Simulation During Processing of a Second Language." This effect of reading in a second language may be stronger with English which has a very "abnormal" structure and vocabulary (see McWhorter, "English Is Not Normal"). Since it is facilitated by adequate emotional response, learning could be made harder for non-native speakers who study in English.

17 Henrich, Heine, and Norenzayan, "The Weirdest People in the World?" This syndrome will be addressed in some detail in chapter 12.

18 Healy, *Endangered Minds: Why Our Children Don't Think,* 222–34. Healy dissected in great detail the philosophy of the program and its likely impact on children's brains. But its strongest impact has probably come from the excuse it has given to parents and educators to expose children from a very young age to an extra one-hour dose of TV viewing on an almost daily basis. Meanwhile, the program has spread like wildfire across the globe, reaching children in more than 140 countries. Of course, exposure to screens playing video has been taken to a whole new level by the explosive spread of tablet computers. In 2013, child protection organizations urged Fischer Price to withdraw a new infant recliner seat designed to hold a tablet in front of a baby's face (see Kang, "Infant iPad Seats Raise Concerns About Screen Time for Babies").

19 A video clip containing fragments from the documentary is available on YouTube. In it, newly minted Harvard graduates claim that it is warmer in summer because the Earth is closer to the sun. Incidentally, a few faculty and staff members

shared the same misconception.

20 See Flesch, *Why Johnny Can't Read—And What You Can Do About It.*

21 National Commission on Excellence in Education, "A Nation at Risk: The Imperatives for Educational Reform." This report was partly motivated by concerns related to the Cold War and the competitiveness of the U.S. economy, but it also reinforced a broader sense that the country's public education system was failing. In 1998, a report drafted under the aegis of the conservative Hoover Institution deemed the nation to be "still at risk" (see Bennett et al., "A Nation Still at Risk"). The No Child Left Behind Act of 2001, the Race to the Top funding contest launched in 2009, the 2012 report "U.S. Education Reform and National Security" of the Council on Foreign Relations task force (chaired by Condoleezza Rice and Joel Klein), the recent Every Student Succeeds Act, and countless other projects for broad educational reform have all reflected similar concerns in the United States—even if the remedies offered may be controversial.

22 See Uddén, Folia, and Petersson, "The Neuropharmacology of Implicit Learning"; Frank, "'Go' and 'noGo': Learning and the Basal Ganglia."

23 See Wilson et al., "Just Think: The Challenges of the Disengaged Mind."

24 See Malamud, "One Big Yawn? The Academics Bewitched by Boredom."

25 Samuelson, "School Reform's Meager Results."

26 See Ravitch, "Schools We Can Envy"; cf. Barkhorn, "America's Math Textbooks Are More Rigorous Than South Korea's."

27 See Porter, "More in School but Not Learning."

28 Quoted in McNeilage, "Students May Be Disadvantaged by Starting School at 5 Years Old"; cf. Whitebread, "Too Much

Too Young: Should Schooling Start at Age 7?"; Christakis, "The New Preschool Is Crushing Kids"; Brosco and Bona, "Changes in Academic Demands and Attention-Deficit/Hyperactivity Disorder in Young Children."

29 See Carr, "Is Google Making Us Stupid?"

11. The Pursuit of Overstimulation

1 See Barrett, *Supernormal Stimuli: How Primary Urges Overran Their Evolutionary Purpose.*

2 See Robbins, "Missing the Big Picture."

3 Quoted in ibid.

4 See Jabr, "How the Brain Ignores Distractions."

5 Quoted in McGowan, "Addiction: Pay Attention."

6 Yoffe, "Seeking."

7 See Chabris, "Is the Brain Good at What It Does?"

8 See Greenfield, *Mind Change*, 209.

9 See Zimbardo and Coulombe, *Man (Dis)connected: How the Digital Age Is Changing Young Men Forever.*

10 See Barrett, *Supernormal Stimuli.*

11 See Moss, "The Extraordinary Science of Addictive Junk Food."

12 Whybrow, *American Mania.*

13 See Linden, *The Compass of Pleasure: How Our Brains Make Fatty Foods, Orgasm, Exercise, Marijuana, Generosity, Vodka, Learning, and Gambling Feel so Good*; Levitin, "Why the Modern World Is Bad for Your Brain"; Hanna, *Stressaholic: 5 Steps to Transform Your Relationship with Stress.*

14 See Previc, *The Dopaminergic Mind in Human Evolution and History.*

15 See Dokoupil, "Is the Internet Making Us Crazy? What the New Research Says."

16 Previc, *Dopaminergic Mind*, 162.

17 See Elsevier Health Sciences, "Am I Fat? Many of Today's Adolescents Don't Think So: Body Weight Perceptions

Changing in the US."

18 See Shin, "4 In 5 Millennials Optimistic for Future, But Half Live Paycheck To Paycheck."

19 See Brooks, "Midlife Crisis Economics."

20 See Kraus, Davidai, and Nussbaum, "American Dream? Or Mirage?"

21 See Whybrow, "Dangerously Addictive."

22 See Brown, *Speed: Facing Our Addiction to Fast and Faster—and Overcoming Our Fear of Slowing Down.*

23 See Marcuse, *Eros and Civilization*; Crawford, "Shop Class as Soulcraft."

24 See Bergen-Cico, "The 'Lee Robins Study' and Its Legacy."

25 See Alexander, "The Myth of Drug-Induced Addiction."

26 See Whybrow, *Well-Tuned Brain*, 21–4; Berreby, "The Obesity Era."

27 See Levitin, "Modern World Is Bad for Your Brain."

28 Quoted de Zengotita, "Numbing of the American Mind."

29 See Simmel, "The Metropolis and Mental Life."

30 McLuhan, *Understanding Media*, 46.

31 See Toffler, *Future Shock*, 350.

32 Milgram, "The Experience of Living in Cities."

33 See Ortega y Gasset, *The Dehumanization of Art, and Other Writings on Art and Culture.*

34 Quoted in Goldman, *The Cambridge Introduction to Virginia Woolf*, 123.

35 The Paris *flâneur* was typically a bohemian artist or writer who would stroll incognito along the streets and arcades of the big city looking primarily for rich sensory experiences. In the words of Evgeny Morozov, "his goal was to observe, to bathe in the crowd, taking in its noises, its chaos, its heterogeneity, its cosmopolitanism"; or "to cultivate what Honoré de Balzac called 'the gastronomy of the eye'." ("The Death of the Cyberflâneur"). This search for novel sensations was probably motivated by a sense of jadedness under

the onslaught of the sensory overstimulation provided by the big city. Once that mild shell-shock passed a certain limit, a leisurely walk could no longer provide the uplift it previously did. The Paris flâneur then disappeared, moving on to search for stronger sensations.

36 See Gergen, *The Saturated Self: Dilemmas of Identity in Contemporary Life.*

37 See Sardamov, "Burnt into the Brain: Towards a Redefinition of Political Culture."

38 See Small and Vorgan, *iBrain.*

39 Tocqueville, *Democracy in America*, vol. 2, ch. ii, "Of Individualism in Democratic Countries."

40 Ibid.

41 Rushkoff, *Life, Inc.: How the World Became a Corporation, and How to Take It Back.*

42 See Goleman, "Feeling Unreal? Many Others Feel the Same"; Simeon and Abugel, *Feeling Unreal: Depersonalization Disorder and the Loss of the Self.*

43 McLuhan, *The Gutenberg Galaxy: The Making of Typographic Man*, 19. McLuhan contrasted literate to non-literate culture, one emphasizing the eye and the other the ear in their perception of the world.

44 See Bergen, *Louder than Words: The New Science of How the Mind Makes Meaning.*

45 McKibben, "The Mental Environment."

46 See Postman, *Teaching as a Conserving Activity*, 186.

12. ...and Existential Disconnect

1 See Flynn, *What Is Intelligence? Beyond the Flynn Effect.* According to some researchers, the Flynn effect may have come to an end, at least in some Western countries (see Sundet, Barlaug, and Torjussen, "The End of the Flynn Effect?".

2 See Case Western Reserve University, "Empathy Represses

Analytic Thought, and Vice Versa: Brain Physiology Limits Simultaneous Use of Both Networks."

3 See Sardamov, "Out of Touch: The Analytic Misconstrual of Social Knowledge."

4 See Lieberman, *Social: Why Our Brains Are Wired to Connect.*

5 See Bergen, *Louder Than Words.*

6 See Lieberman, *Social*; Adolphs, "The Social Brain: Neural Basis of Social Knowledge"; Mars et al., "On the Relationship Between the 'Default Mode Network' and the 'Social Brain.'"

7 See Jack et al., "fMRI Reveals Reciprocal Inhibition Between Social and Physical Cognitive Domains."

8 Jack, "A Scientific Case for Conceptual Dualism: The Problem of Consciousness and the Opposing Domains Hypothesis," 181 (emphasis in original).

9 See Adolphs, "Social Brain."

10 See Yang and Li, "Brain Networks of Explicit and Implicit Learning"; Paul, "Learning but Not Trying."

11 Mars et al., "'Default Mode Network' and 'Social Brain,'" 7.

12 See Kaufman, "Social and Mechanical Reasoning Inhibit Each Other."

13 See Boyle, *The Tyranny of Numbers: Why Counting Can't Make Us Happy*, xiii.

14 See Ratey and Johnson, *Shadow Syndromes: The Mild Forms of Major Mental Disorders That Sabotage Us.*

15 See Carson, "Creativity and Psychopathology: A Shared Vulnerability Model."

16 See Gladwell, *Outliers: The Story of Success*, 69–115.

17 Above a certain level of complexity, mathematics may in fact require a more holistic perspective—which could account for the number of gifted mathematicians who have developed psychotic (rather than autistic) symptoms or even committed suicide as they have tackled baffling mathematical puzzles (see Wallace, *Everything and More: A Compact*

History of Infinity).

18 Neuroscientists have uncovered some curious structural differences between the brains of science and humanities students, with the former displaying some quasi-autistic traits (see Takeuchi et al., "Brain Structures in the Sciences and Humanities"; cf. Tanaka, "Autism, Psychosis, and the 'Two Cultures'"). Needless to say, such a divergence can only be reinforced as the two groups engage in their different fields of study and exploration.

19 This more specific condition was removed from DSM-5 as it was subsumed under the broader diagnosis of "autism spectrum disorder." Individuals are allowed to retain the old diagnosis if, for some reason, they wish to do so.

20 See Case Western Reserve University, "Empathy Represses Analytic Thought."

21 Ibid.

22 See Polanyi, *The Great Transformation*, 111–16.

23 Psychologist James Flynn has attributed much of the increase in IQ scores associated with his name to increased level of complexity and stimulation in modern societies (see Flynn, *What Is Intelligence?*).

24 The evaluations students completed in April 2014 for my upper level course, "Culture and Power," displayed a particularly troubling tendency. A few students complained that the question marks which I had placed at the end of some statements or phrases in the PowerPoint outlines I showed in class were confusing (though these were only meant as a reminder that some complex problems defy a definite explanation or solution). There were also several complaints (by the same or other students) that they had become lost during the last few weeks of the semester, though I had repeatedly tried to outline the overall logic of the course.

25 See Immordino-Yang, Christodoulou, and Singh, "Rest Is Not Idleness: Implications of the Brain's Default Mode for

Human Development and Education."

26 See Mars et al., "'Default Mode Network' and 'Social Brain.'"

27 See Chiao and Immordino-Yang, "Modularity and the Cultural Mind: Contributions of Cultural Neuroscience to Cognitive Theory."

28 See Pappas, "When You're at Rest, Your Brain's Right Side Hums."

29 See Adolphs, "Social Brain"; Sridharan, Levitin, and Menon, "A Critical Role for the Right Fronto-Insular Cortex in Switching Between Central-Executive and Default-Mode Networks."

30 See Heberlein et al., "Effects of Damage to Right-Hemisphere Brain Structures on Spontaneous Emotional and Social Judgments"; Lewis, "Think You Have Self-Control? Careful."

31 Jack, personal communication, Nov. 26, 2013

32 See Wanjek, "Left Brain vs. Right: It's a Myth, Research Finds"; cf. McGilchrist, *The Master and His Emissary: The Divided Brain and the Making of the Western World.*

33 See Brooks, "The Empirical Kids."

34 See Yang and Li, "Brain Networks of Explicit and Implicit Learning"; Paul, "Learning but Not Trying."

35 See Richtel, "Digital Devices Deprive Brain of Needed Downtime."

36 During AUBG's 2015 commencement ceremony, there were five salutatorians on the stage. Two of them had been demoted to this status because of an A- they had received from me in their senior year. They had both suffered this iniquity because they were, in my judgment, unable to produce a well written paragraph and make some broader associations. Yet, they both believed they fully deserved to be valedictorians. Again, my consolation was that I am not alone facing similar problems (see Schuman, "Confessions of a Grade Inflator"). Incidentally, such cases serve to

reinforce my gratitude to those truly exceptional students who combine a wide-eyed curiosity with a "growth mindset" and ability to recognize areas for improvement (see Dweck, *Mindset: The New Psychology of Success*).

37 Most of these test subjects had in fact been American undergraduates who appeared to be the most extreme outliers, the weirdest of the weird (see Henrich, Heine, and Norenzayan, "Weirdest People").

38 See Nisbett, *The Geography of Thought: How Asians and Westerners Think Differently … and Why*.

39 Ibid.

40 See Han and Northoff, "Understanding the Self: A Cultural Neuroscience Approach."

41 See Polanyi, *Great Transformation*.

42 There is research indicating that involuntary blinking (which is normally more frequent than what is necessary for eye lubrication) is associated with momentary attentional disengagement and activation of the default mode network (see Nakano et al., "Blink-Related Momentary Activation of the Default Mode Network While Viewing Videos"). Incidentally, blinking is less frequent when we interact with screens. Since the intuitive network is so central to the overall integration of the brain and of the brain and body, it has apparently evolved to be activated during every spare moment in our daily lives (see Lieberman, *Social*, 19-20).

43 See Brooks, "Empirical Kids."

44 Incidentally, these predispositions lead most economically unsuccessful Americans to blame mostly themselves for their misfortune (see Phillips-Fein, "Why Workers Won't Unite").

45 According to some researchers, the steady increase in IQ scores which started over a century ago does not reflect gains in "general intelligence." Rather, it results from increasing aptitude for taking all kinds of tests and solving rule-dependent problems (see Robb, "Our IQs Are Climbing, But

We're Not Getting Smarter").

46 See Sardamov, "Out of Touch."

47 Watters, "We Aren't the World."

48 A prime example is the predominant instrumentalist approach to ethnic nationalism which focuses strongly on manipulation by unscrupulous leaders who pursue power and control over resources. Typically, such explanations of ethnic conflict blame large-scale violence on the likes of Serbian president Slobodan Milosevic (see Sardamov, "Mandate of History: Serbian National Identity and Ethnic Conflict in the Former Yugoslavia"). Islamist terrorism seems less amenable to similar reductive explanations, but the alternatives are too disturbing to contemplate.

49 British author and journalist David Boyle provides a particularly evocative portrayal of Jeremy Bentham and Edwin Chadwick who in the 19th century exemplified these tendencies—applying much mental resourcefulness and self-assurance in the obsessive, somewhat childish pursuit of various utilitarian projects (see Boyle, *Tyranny of Numbers*, 16–35, 65–82). In recent years, famed psychologist Steven Pinker has offered the most influential version of "Whig history" positing progress toward stronger empathy, mutual understanding, and the consequent waning of violence on a global scale(see Pinker, *The Better Angels of Our Nature: Why Violence Has Declined*).

13. Unenlightened

1 Updike, "Beer Can."

2 See Dunckley, "The Hidden Agenda Behind 21st Century Learning."

3 See Willis, "A Neurologist Makes the Case for the Video Game Model as a Learning Tool." This article is posted on Edutopia.org—a website maintained by the George Lucas Educational Foundation, an organization dedicated to

"brain-based learning."

4 See Oppenheimer, "The Computer Delusion."

5 Willis, *Research-Based Strategies to Reignite Student Learning.*

6 See Arrowsmith-Young and Doidge, *The Woman Who Changed Her Brain: How I Left My Learning Disability Behind and Other Stories of Cognitive Transformation.*

7 Luria's research is described in Crain, "Twilight of Books."

8 See Wolf, *Proust and the Squid: The Story and Science of the Reading Brain*; cf. Dehaene, *Reading in the Brain: The Science and Evolution of Human Invention.*

9 See Birkerts, *Gutenberg Elegies*; cf. Wolf and Barzillai, "The Importance of Deep Reading."

10 "Deep reading" is often confused with "close reading"—the detailed analysis of literary passages long favored at college English departments. According to poet and writer Dana Gioia, such nitpicking textual analysis stands in the way of truly appreciating poetry—and perhaps this verdict holds for literature in general (see Gioia, "Poetry as Enchantment"). Some experts nevertheless believe that instruction in "close reading" is essential for helping children comprehend what they read, and cannot wait until middle school (see Boyles, "Closing in on Close Reading").

11 See Coyle, *The Talent Code: Greatness Isn't Born. It's Grown*; Gladwell, *Outliers*. The 10,000-hour rule was first formulated in 1993 by psychologist K. Anders Ericsson, but has been recently questioned as reductionist (see Carey, "How Do You Get to Carnegie Hall? Talent.").

12 Myelination is an exceedingly intricate process which can easily be disrupted by chronic stress (see University at Buffalo, "New Form of Brain Plasticity: How Social Isolation Disrupts Myelin Production"). It also needs to be supplemented by proper pruning of inessential synaptic connections during the adolescent years—a similarly complex and tentative developmental process which can go awry, causing

developmental disruptions (see Siegel, *Brainstorm: The Power and Purpose of the Teenage Brain*, 97–100).

13 Darnton, "The First Steps toward a History of Reading," quoted in Birkerts, *Gutenberg Elegies*, 71.

14 Idem, 72.

15 See Carr, "Is Google Making Us Stupid?'; *Shallows*.

16 Quoted in Carr, *Shallows*, 111.

17 Curiously, Shirky eventually decided to ban laptops and other electronic devices in the classroom as he has observed growing levels of distraction among his students—not just in those using the equipment, but also among students sitting nearby (see Shirky, "Why Clay Shirky Decided to Ban Laptops, Tablets, and Phones from His Classroom").

18 Burns, "Dopamine and Learning: What the Brain's Reward Center Can Teach Educators" (emphasis in original).

19 Back in 2010, Harvard started offering a highly popular course (now available online) on "science and cooking" taught with the help of a few celebrity chefs. It falls within a broader thrust (starting much before college) aimed to make basic scientific concepts and theories more relevant and interesting—for example, by introducing their application in forensic science or other exciting areas.

20 Willis, *Research-Based Strategies to Reignite Student Learning*, 59; cf. Willingham, *Why Don't Students Like School? A Cognitive Scientist Answers Questions About How the Mind Works and What It Means for the Classroom*; Sousa, *Mind, Brain, and Education*.

21 In late 2014, the MIT cut all ties with Walter Lewin, who had retired in 2009. This decision was taken after an investigation established that he had sexually harassed online at least one student (who was taking one of his courses on edX). Prof. Lewin's lectures are still available online, though not on MIT-affiliated websites.

22 See Khadaroo, "When College Students Reinvent the

World."

23 Edmundson, "Dwelling on Possibilities."

24 See Lambert, "Nonstop: Today's Undergraduates Schedule Themselves 24/7. But Are They Missing Opportunities for Self-Discovery?"

25 Ibid.; cf. Brooks, "Organization Kid."

26 See Haiken, "Lack of Sleep Kills Brain Cells, New Study Shows"; Cedernaes et al. "Acute Sleep Loss Induces Tissue-Specific Epigenetic and Transcriptional Alterations to Circadian Clock Genes in Men."

27 Davidson, *Now You See It: How the Brain Science of Attention Will Transform the Way We Live, Work, and Learn.*

28 This experiment is also open to a different interpretation—as an illustration of the ability of the overly focused and analytic mind (and brain) to filter out potentially significant background information. But Davidson would probably be unconvinced.

29 Ibid., 278, 280–1.

30 Ibid., 285. Heavy media multitaskers may, in fact, perform better in "noisy" environments—but have difficulty filtering out irrelevant stimuli when they need to focus on a task, and switch less efficiently between tasks (see Ophir, Nass, and Wagner, "Cognitive Control in Media Multitaskers").

14. Toward Edutopia?

1 See National Association for the Education of Young Children, "Play and Children's Learning." Unfortunately, none of the eight boys from our apartment building I regularly played with went on to pursue a higher education degree.

2 Birkerts, an even more avid reader, has similarly noted: "I value the state a book puts me in more than I value the specific contents" (*Gutenberg Elegies*, 85). Perhaps voracious reading can also be seen as a form of overindulgence, or even

addiction, which could get in the way of directly experiencing reality (see Robson, "Living Life by the Book: Why Reading Isn't Always Good for You"). Alas, we may be at a point where escaping all such addictive behaviors is no longer a realistic option—and deep reading is perhaps the healthiest mental preoccupation (unless it leads to an overly sedentary lifestyle and chronic sleep deprivation).

3 See Reynolds, "What You Read Matters More Than You Might Think."

4 See Willingham, *Why Don't Students Like School?*, 49.

5 Cunningham and Stanovich, "What Reading Does to the Mind."

6 See Willingham, "How Knowledge Helps: It Speeds and Strengthens Reading Comprehension, Learning—and Thinking."

7 See Crain, "Twilight of Books."

8 These issues are systematically addressed in the *Moral Landscapes* blog maintained by developmental psychologist Darcia Narvaez on the *Psychology Today* website.

9 See Steiner-Adair, *The Big Disconnect: Protecting Childhood and Family Relationships in the Digital Age*; cf. Fallows, "Papa, Don't Text: The Perils of Distracted Parenting."

10 See Washbrook, Waldfogel, and Moullin, *Parenting, Attachment and a Secure Base for Children.*

11 See Narvaez, "How to Grow a Smart Baby."

12 See Schulte, "Effects of Child Abuse Can Last a Lifetime: Watch the 'Still Face' Experiment to See Why."

13 See *BMJ*, "2-Year-Olds Adept at Using Touch-Screen Technology: Able to Swipe, Unlock, and Actively Search for Features on Phones and Tablets."

14 Psychologist Matthew Lieberman has argued that our "social brain" has a natural appetite for social connection which will always assert itself (see Lieberman, *Social*). As various forms of addiction illustrate, though, healthy

natural instincts (and the neurosomatic mechanisms underlying these) can sometimes be hijacked by various forms of overindulgence.

15 The experience of the children of deaf parents is very instructive in this respect. Such children could potentially learn to speak, enlarge their vocabulary, and achieve a sense of grammar and syntax from television. In reality, they often have serious linguistic impairments when deprived of sufficient live communication (see Byeon and Hong, "Relationship Between Television Viewing and Language Delay in Toddlers: Evidence from a Korea National Cross-Sectional Survey").

16 See Healy, *Endangered Minds*, 218–34.

17 See Dunckley, *Reset Your Child's Brain*, 42–4.

18 The American Academy of Pediatrics recommends that pediatricians advise parents about the benefits of reading aloud to young children and talk with them about age-appropriate stories and pictures. According to some studies, digital books can interfere with such interactions, as parents and children tend to focus on the electronic device itself (see Quenqua, "Is E-Reading to Your Toddler Story Time, or Simply Screen Time?").

19 Klass, "Bedtime Stories for Young Brains."

20 See Bodrova, Germeroth, and Leong, "Play and Self-Regulation: Lessons from Vigotsky"; Lillard et al., "The Impact of Pretend Play on Children's Development: A Review of the Evidence." A recent study has found that a decrease in physical activity and motor skill has resulted in more frequent injuries in children (see Nauta et al.,"A Systematic Review on the Effectiveness of School and Community-Based Injury Prevention Programmes on Risk Behaviour and Injury Risk in 8–12 Year Old Children").

21 See Edwards, "Three Approaches from Europe: Waldorf, Montessori, and Reggio Emilia"; Oppenheimer, "Schooling

the Imagination." According to psychologists Kevin Rathunde and Mihaly Csikszentmihalyi, American students in Montessori middle schools tend to have stronger intrinsic motivation to learn, and report more absorbed engagement in schoolwork (or "flow"—see Rathunde and Csikszentmihalyi, "Middle School Student' Motivation and Quality of Experience: A Comparison of Montessori and Traditional School Environments").

22 See Zernike, "Fast-Tracking to Kindergarten?"; Christakis, "New Preschool Is Crushing Kids."

23 Hamilton, "Reading Practice Can Strengthen Brain 'Highways"; Cunningham and Stanovich, "What Reading Does to the Mind."

24 Healy, *Endangered Minds*, 89.

25 Some children who are developmentally precocious and eager can learn to read and write earlier, a practice encouraged in Montessori schools through individualized instruction. Such earlier cognitive development, however, should not be forced.

26 Alas, such a reading-focused learning environment may not be beneficial for students with strong attention deficit or dyslexia symptoms. Implementing it may therefore require streaming students on the basis or their potential to become earnest readers.

27 See Bounds, "How Handwriting Trains the Brain: Forming Letters Is Key to Learning, Memory, Ideas"; Klass, "Why Handwriting Is Still Essential in the Keyboard Age." Unfortunately, instruction in cursive handwriting is being phased out after the first few grades of schooling in the United States and many other countries in favor of typing—which is seen as a 21st century skill.

28 See May, "A Learning Secret: Don't Take Notes with a Laptop."

29 Healy, *Endangered Minds*, 170; cf. Belluck, "Traditional Toys

May Beat Gadgets in Language Development."

30 See Murgia, "Technology in Classrooms Doesn't Make Students Smarter."

31 See Baron, *Words Onscreen: The Fate of Reading in a Digital World*.

32 Baron, "How E-Reading Threatens Learning in the Humanities."

33 See Richtel, "A Silicon Valley School That Doesn't Compute."

34 See Hirsch, *The Making of Americans: Democracy and Our Schools*; Brown, Roediger III, and McDaniel, *Make It Stick*.

35 Quoted in Durayappah-Harrison, "The Secret Benefits of a Curious Mind."

36 See Kafka, "From the Desk of Roland Barthes."

37 See Kolowich, "Confuse Students to Help Them Learn."

38 See Lieberman, *Social*, 275–98.

39 According to a curious recent study, medical students learn anatomy better when the instruction process uses old-fashioned cadavers as opposed to cutting-edge computer simulations. Apparently, the "real thing" makes a stronger impression on most students (see Saltarelli, Roseth, and Saltarelli. "Human Cadavers vs. Multimedia Simulation: A Study of Student Learning in Anatomy").

40 Claxton, *Intelligence in the Flesh: Why Your Mind Needs Your Body Much More Than It Thinks*, 272.

41 Barber, "Shrunken Sovereign: Consumerism, Globalization, and American Emptiness." The series of articles on the American "disunion" 150 years ago in *The New York Times* provides another good example (http://opinionator.blogs.nytimes.com/category/disunion).

42 See Dalrymple, "The Case for Cannibalism."

43 See Durham University, "Star Trek Classroom: Next Generation of School Desks."

44 This tendency is perhaps illustrated by a simulation exercise I observed in November 2013 at our department. It involved

stylized international negotiations during which teams representing eurozone countries were able to score points as they adopted different policies. Some of the students were able to design a strategy which, once adopted by all teams, ensured that they all finish with an equal number of points, and all receive the bonus points that were to be allocated to the winners. This outcome produced much excitement and cheering. I had a sense that the build-up to it and the exhilaration it produced were much more likely to stick in most student's minds than any substantive lessons the negotiation simulation was supposed to teach them.

45 Healy, *Endangered Minds,* 214; Lehrer, "The Educational Benefits of Ugly Fonts."

46 Doidge, "Building a Better Brain." There is also some research indicating that boys struggling in school may benefit from a more authoritative teaching style (see Hadjar, Backes, and Gysin, "School Alienation, Patriarchal Gender-Role Orientations and the Lower Educational Success of Boys: A Mixed-Method Study").

47 See Paul, "Why Schools' Efforts to Block the Internet Are So Laughably Lame, and Why It's Seriously Important to Keep Students Off Facebook."

48 See McLeod, "Jean Piaget."

49 There is recent research indicating that even excessive decoration in kindergarten classrooms (much of which is directly intended to provide visual props for learning) can be counterproductive (see Hoffman, "Rethinking the Colorful Kindergarten Classroom").

50 See Christakis, "New Preschool Is Crushing Kids."

51 See Jabr, "The Reading Brain in the Digital Age: Why Paper Still Beats Screens." If the meaning of a text derives mostly from our emotional responses to it, and reading from a screen evokes a denuded emotional response, then (and even some non-fiction) will have a different meaning when

"delivered" through different media. In a sense, reading the same text from a physical "codex" (arranged collection of pages) and from a screen will be like reading two slightly different texts.

52 See Dartmouth College. "Digital Media May Be Changing How You Think: New Study Finds Users Focus on Concrete Details Rather Than the Big Picture."

53 See Paul, "Reading Literature Makes Us Smarter and Nicer."

54 See Carr, "As Technology Advances, Deep Reading Suffers"; Jabr, "Reading Brain in the Digital Age."; Baron, *Words Onscreen*.

55 See Giraldi, "Object Lesson: Why We Need Physical Books."

56 Quoted in Bauerlein, "Screen Reading and Print Reading."

57 See Millward Brown, "Using Neuroscience to Understand the Role of Direct Mail."

58 Clabby, "Brains Like to Keep It Real."

59 See Arum and Roksa, *Academically Adrift*.

60 See Brooks, "Empirical Kids"; Sardamov, "Out of Touch."

61 See Willingham, *Why Don't Students Like School?*, 98–9.

62 See Schroeder, "New Students—New Learning Styles."

63 See Cohen, "An SAT Without Analogies Is Like: (A) A Confused Citizenry…"

64 See Arum and Roksa, *Academically Adrift*.

65 See Haidt, *Righteous Mind*; Jack, "Conceptual Dualism."

66 See Cunningham and Stanovich, "What Reading Does to the Mind."

67 See Dunckley, *Reset Your Child's Brain*.

68 See Arum and Roksa, *Academically Adrift*.

69 See Everett, "Cultural Constraints on Grammar and Cognition in Pirahã: Another Look at the Design Features of Human Language."

70 See Willingham, *Why Don't Students Like School?*

15. The Last Oasis?

1 See Whybrow, *Well-Tuned Brain*, 293–312; Previc, *Dopaminergic Mind*, 155–72.
2 Whybrow, idem, xiii.
3 Idem, 176–209.
4 Idem, 206. Whybrow refers to "this collection of skills for resilience, empathic understanding, self-control, and prudent planning as *character*" (ibid., emphasis in original).
5 See Hanna, *Stressaholic*; Sieberg, *The Digital Diet: The 4-Step Plan to Break Your Tech Addiction and Regain Balance in Your Life*; Linden, *Compass of Pleasure*; Levitin, *The Organized Mind: Thinking Straight in the Age of Information Overload*.
6 See Zimbardo and Coulombe, *Man (Dis)connected*.
7 See Greenfield, *Mind Change*, 274–86; cf. McKibben, "Mental Environment."
8 Bell, Bishop, and Przybylski, "The Debate Over Digital Technology and Young People."
9 Bell, "Head in the Cloud"; cf. Neuroskeptic, "Susan Greenfield Causes Autism"; Goldacre, "Speculation, Hypothesis and Ideas. But Where Is the Evidence?"
10 See Bolton, "Smartphones Are Making Children Borderline Autistic, Says Psychiatrist"; Oxenham, "No, Children Aren't Making Children Autistic."
11 My general sense is that our beliefs about the world, people, and ideas can tell us more about the way we function neurophysiologically as opposed to the objects we contemplate (see Sardamov, "Out of Touch"). Which does not mean all beliefs are created equal, and everything is relative and potentially plausible (or less so). I tend to trust thinkers who seem to combine obvious intelligence and sufficient background knowledge with keen emotional and interoceptive sensitivity (see note 13 below).
12 See Greenfield, *Mind Change*, 30.
13 There are various way to assess someone's affective-visceral

reactivity—from brain scanning to measuring electrical skin conductance (as done in a polygraph test), or simply testing the ability of a person to sense their own heart rate at rest. As neuroscientist Hugo Critchley points out, "the better you are at tracking your own heartbeats, the better you are at experiencing the full gamut of human emotions and feelings. The more viscerally aware, the more emotionally attuned you are" (quoted in Blakeslee and Blakeslee, "My Insula Made Me Do It"). There is, indeed, growing recognition that such deeper affective-visceral attunement is key to adequate social understanding and judgment (and it tends to correlate with holistic thinking and more apprehensive attitudes). In the absence of data regarding the interoceptive sensitivity of intellectual opponents, I tend to rely on my own gut feeling and holistic assessment—which tell me I should rather trust the Baroness than her self-assured critics (for a fuller elaboration of this epistemological position, see Sardamov, "Out of Touch").

14 Goldberg, *Executive Brain*, 222.

15 Idem, 219–23.

16 Goldberg, *New Executive Brain*, 275–81. Goldberg has also dismissed anxieties about an impending "anarchy" in the digital world—suggesting that search engines have effectively become "the *digital frontal lobes*" (281, emphasis in original).

17 Greenberg, "Why Last Chapters Disappoint."

18 Ibid.

19 See Watters, "We Aren't the World"; Scicurious, "Dopamine Goggles Make the Glass Half-Full"; Chang and Asakawa, "Cultural Variations on Optimistic and Pessimistic Bias for Self Versus a Sibling: Is There Evidence for Self-enhancement in the West and for Self-criticism in the East when the Referent Group Is Specified?"

20 See Eyal and Hoover, *Hooked: How to Build Habit-Forming*

Products; Zuboff, "The Secret of Surveillance Capitalism."

21 See American Educational Research Association, "Little Evidence That Executive Function Interventions Boost Student Achievement"; cf. Bornstein, "Teaching Social Skills to Improve Grades and Lives." Disappointing results from programs emphasizing non-cognitive skill may partly reflect a negative effect of increased executive control on implicit learning.

22 See Abeles, "Is the Drive for Success Making Our Children Sick?" Leaving behind her career as a Wall Street attorney, in 2009 Vicki Abeles produced and co-directed the documentary *Race to Nowhere* highlighting these dangers. She later made another documentary, *Beyond Measure,* exploring innovative educational approaches.

23 See Previc, *Dopaminergic Mind,* 75.

24 See Reed, "Study: Digital Distraction in Class Is on the Rise."

25 Natural birth is associated with healthier growth and development in babies (see Block, "Our C-Section Rate Won't Bulge—Is It Because We Don't Trust Women's Hormones?").

26 The World Health Organization recommends breastfeeding for up to two years, exclusively for the first six months of life (World Health Organization, "Exclusive Breastfeeding").

27 Maté, *In the Realm of Hungry Ghosts: Close Encounters with Addiction,* 240; see 240-59.

28 Palmer, *Toxic Childhood: How the Modern World Is Damaging Our Children and What We Can Do About It.*

29 See Aron, *The Highly Sensitive Child: Helping Our Children Thrive When the World Overwhelms Them.* This neurosomatic type is akin to the "orchid children" and those "living with intensity" and heightened excitabilities mentioned earlier (see Dobbs, "Science of Success"; Daniels, "Living with Intensity: Overexcitabilities in Profoundly Gifted Children"). Psychologist Jerome Kagan has described in less positive light the typical temperament of high-reactive

children (*The Long Shadow of Temperament*).

30 See Dobbs, ibid.; "Can Genes Send You High or Low? The Orchid Hypothesis A-Bloom"; Bartz, "Sense and Sensitivity."

31 See Hayles, "Hyper and Deep Attention: The Generational Divide in Cognitive Modes."

32 Ibid.

33 See Herbert and Weintraub, *Autism Revolution*; Miller, "Is the World Making You Sick?"; Chang, "Debate Continues on Hazards of Electromagnetic Fields."

34 See Galambos, Barker, and Tilton-Weaver, "Who Gets Caught at Maturity Gap? A Study of Pseudomature, Immature, and Mature Adolescents."

35 See Elias, *The Civilizing Process*. Curiously, Pinker believes American society has since the 1990s experienced a "recivilizing process," reversing the decivilizing trend dating back to the 1960s. He marshals much data showing decreasing levels of violent crime and other forms or symptoms of social pathology (see Pinker, *Better Angels*, 116–28).

36 See Sardamov, "From 'Bio-Power' to 'Neuropolitics': Stepping Beyond Foucault."

37 Stiegler, *Taking Care of Youth and the Generations*, 2 (emphasis in original).

38 Stiegler does place some hope in recent technological trends like the development of open-source and free software, and of collaborative practices and communities making use of these (idem).

39 It is well known that video games and websites are designed to be addictive, so they can attract the most clicks and elicit the longest "user experience" possible. A young Facebook "research scientist" commented ruefully a few years ago: "The best minds of my generation are thinking about how to make people click ads. That sucks." (quoted in Vance, "This Tech Bubble Is Different").

40 Birkerts, *Gutenberg Elegies*, 190.

41 See Beck, *Risk Society: Towards a New Modernity*.

42 This is my own loose translations of the slogan which comes from *The Twelve Chairs*, a much beloved satirical novel written by Ilya Ilf and Evgeny Petrov.

43 See Neufeld and Maté, *Hold On to Your Kids: Why Parents Need to Matter More Than Peers* (the updated edition of the book includes two chapters on the challenges of parenting in the digital age). According to one recent study, sixth-graders who spent just five days at a camp which bans TV and all digital devices showed a significant improvement in their ability to relate emotionally to others—and parents could similarly benefit from reduced screen time (see Wolpert, "In Our Digital World, are Young People Losing the Ability to Read Emotions?".

44 When Gali was 12 years old, we were once standing in the street, waiting for a bus under the falling snow. As we were chatting, she suddenly exclaimed, "Look, it is exactly the way they draw it!" She was looking at a single snowflake sitting on the sleeve of my dark woolen coat.

45 See Lewis, "The End."

46 Bauerlein, *Dumbest Generation*.

47 See Previc, *Dopaminergic Mind*, 163–4.

48 Duff, "The Information Revolution's Dark Turn."

49 See Popper, "The Robots Are Coming for Wall Street."

50 See Huxley, *Brave New World*.

51 Mendelson, "In the Depth of the Digital Age."

52 Postman, *Amusing Ourselves to Death*, viii.

53 Ibid.

54 See McGowan, "Addiction."

55 Maté, *Hungry Ghosts*, 440.

56 This happened in 2010 at Cape Denison in Antarctica. The penguins then needed to waddle 60 km (over 37 miles) to catch fish. In five years, their colony shrunk by 150,000—with the remaining 10,000 apparently facing a dire future.

57 See Jenouvrier et al., "Effects of Climate Change on an Emperor Penguin Population: Analysis of Coupled Demographic and Climate Models." Of course, penguins are not compounding their own predicament the way we, humans are.

58 Renowned social critic Lewis Mumford once criticized a type of liberals overly attached to "rational science and experimental practice." In his view, such a liberal is always "hoping for the best" while he "remains unprepared to face the worst; and on the brink of what may turn out another Dark Age, ... continues to scan the horizon for signs of dawn" ("The Corruption of Liberalism"). Writing at the outset of World War II, Mumford attributed this Pollyanna optimism to lack of adequate emotional response in the face of a mortal threat.

Bibliography

The complete bibliography is available online at https://goo.gl/eTwpcU)

Iff Books

ACADEMIC AND SPECIALIST

Iff Books publishes non-fiction. It aims to work with authors and titles that augment our understanding of the human condition, society and civilisation, and the world or universe in which we live.
If you have enjoyed this book, why not tell other readers by posting a review on your preferred book site. Recent bestsellers from Iff Books are:

Why Materialism Is Baloney
How True Skeptics Know There is no Death and Fathom Answers to Life, the Universe, and Everything
Bernardo Kastrup
A hard-nosed, logical, and skeptic non-materialist metaphysics, according to which the body is in mind, not mind in the body.
Paperback: 978-1-78279-362-5 ebook: 978-1-78279-361-8

The Fall
Steve Taylor
The Fall discusses human achievement versus the issues of war, patriarchy and social inequality.
Paperback: 978-1-90504-720-8 ebook: 978-184694-633-2

Brief Peeks Beyond

Critical Essays on Metaphysics, Neuroscience, Free Will, Skepticism and Culture

Bernardo Kastrup

An incisive, original, compelling alternative to current mainstream cultural views and assumptions.

Paperback: 978-1-78535-018-4 ebook: 978-1-78535-019-1

Framespotting

Changing How You Look at Things Changes How You See Them

Laurence & Alison Matthews

A punchy, upbeat guide to framespotting. Spot deceptions and hidden assumptions; swap growth for growing up. See and be free.

Paperback: 978-1-78279-689-3 ebook: 978-1-78279-822-4

Is There an Afterlife?

David Fontana

Is there an Afterlife? If so what is it like? How do Western ideas of the afterlife compare with Eastern? David Fontana presents the historical and contemporary evidence for survival of physical death.

Paperback: 978-1-90381-690-5

Nothing Matters

A Book About Nothing

Ronald Green

Thinking about Nothing opens the world to everything by illuminating new angles to old problems and stimulating new ways of thinking.

Paperback: 978-1-84694-707-0 ebook: 978-1-78099-016-3

Panpsychism

The Philosophy of the Sensuous Cosmos

Peter Ells

Are free will and mind chimeras? This book, anti-materialistic but respecting science, answers: No! Mind is foundational to all existence.

Paperback: 978-1-84694-505-2 ebook: 978-1-78099-018-7

Punk Science

Inside the Mind of God

Manjir Samanta-Laughton

Many have experienced unexplainable phenomena; God, psychic abilities, extraordinary healing and angelic encounters. Can cutting-edge science actually explain phenomena previously thought of as 'paranormal'?

Paperback: 978-1-90504-793-2

The Vagabond Spirit of Poetry

Edward Clarke

Spend time with the wisest poets of the modern age and of the past, and let Edward Clarke remind you of the importance of poetry in our industrialized world.

Paperback: 978-1-78279-370-0 ebook: 978-1-78279-369-4

Readers of ebooks can buy or view any of these bestsellers by clicking on the live link in the title. Most titles are published in paperback and as an ebook. Paperbacks are available in traditional bookshops. Both print and ebook formats are available online.

Find more titles and sign up to our readers' newsletter at http://www.johnhuntpublishing.com/non-fiction

Follow us on Facebook at https://www.facebook.com/JHPNonFiction and Twitter at https://twitter.com/JHPNonFiction